Advances in Transport Phenomena

Volume 3

For further volumes:
http://www.springer.com/series/8203

About this Series

The term transport phenomena is used to describe processes in which mass, momentum, energy and entropy move about in matter. Advances in Transport Phenomena provide state-of-the-art expositions of major advances by theoretical, numerical and experimental studies from a molecular, microscopic, mesoscopic, macroscopic or megascopic point of view across the spectrum of transport phenomena, from scientific enquiries to practical applications. The annual review series intends to fill the information gap between regularly published journals and university-level textbooks by providing in-depth review articles over a broader scope than in journals. The authoritative articles, contributed by internationally-leading scientists and practitioners, establish the state of the art, disseminate the latest research discoveries, serve as a central source of reference for fundamentals and applications of transport phenomena, and provide potential textbooks to senior undergraduate and graduate students.

Liqiu Wang
Editor

Advances in Transport Phenomena 2011

Springer

Editor
Liqiu Wang
Department of Mechanical Engineering
The University of Hong Kong
Hong Kong
People's Republic of China

ISSN 1868-8853 ISSN 1868-8861 (electronic)
ISBN 978-3-319-01792-1 ISBN 978-3-319-01793-8 (eBook)
DOI 10.1007/978-3-319-01793-8
Springer Cham Heidelberg New York Dordrecht London

Library of Congress Control Number: 2011923889

Printed on acid-free paper

Springer is part of Springer Science+Business Media (www.springer.com)

Preface

The term *Transport Phenomena* is used to describe processes in which mass, momentum, energy, and entropy move about in matter. *Advances in Transport Phenomena* provide state-of-the-art expositions of major advances by theoretical, numerical, and experimental studies from a molecular, microscopic, mesoscopic, macroscopic, or megascopic point of view across the spectrum of transport phenomena, from scientific enquiries to practical applications. The annual review series intends to fill the information gap between regularly published journals and university-level textbooks by providing in-depth review articles over a broader scope than in journals. The authoritative articles, contributed by internationally leading scientists and practitioners, establish the state of the art, disseminate the latest research discoveries, serve as a central source of reference for fundamentals and applications of transport phenomena, and provide potential textbooks to senior undergraduate and graduate students.

The series covers mass transfer, fluid mechanics, heat transfer, and thermodynamics. The 2011 volume contains three articles on the microfluidic fabrication of vesicles, the dielectrophoresis field-flow fractionation for continuous-flow separation of particles and cells in microfluidic device, and the thermodynamic analysis and optimization design of heat exchanger, respectively. The editorial board expresses its appreciation to the contributing authors and reviewers who have maintained the standard associated with *Advances in Transport Phenomena*. We also would like to acknowledge the efforts of the staff at Springer who have made the professional and attractive presentation of the volume.

Contents

Contributors

Jiangfeng Guo School of Civil Engineering, Institute of Engineering Mechanics, Shandong University, Jinan 250061, People's Republic of China

Shin-Hyun Kim School of Engineering and Applied Sciences and Department of Physics, Harvard University, Cambridge, MA, USA

Nuttawut Lewpiriyawong School of Mechanical and Aerospace Engineering, Nanyang Technological University, Singapore 639798, Singapore

Xuefang Li School of Civil Engineering, Institute of Engineering Mechanics, Shandong University, Jinan 250061, People's Republic of China

Ho Cheung Shum Department of Mechanical Engineering, University of Hong Kong, Pokfulam Road, Hong Kong, People's Republic of China

Julian Thiele Institute for Molecules and Materials, Radboud University Nijmegen, Heyendaalseweg 135, 6525 AJ Nijmegen, The Netherlands

Mingtian Xu School of Civil Engineering, Institute of Engineering Mechanics, Shandong University, Jinan 250061, People's Republic of China

Chun Yang School of Mechanical and Aerospace Engineering, Nanyang Technological University, Singapore 639798, Singapore

Microfluidic Fabrication of Vesicles

Ho Cheung Shum, Julian Thiele and Shin-Hyun Kim

Abstract Vesicles are compartments enclosed by a thin membrane, which is made up of amphiphilic molecules arranged into ordered layers. Vesicle-like structures are Nature's choice for encapsulating important biochemical species that enable living processes, and are increasingly important as artificial structures for the encapsulation and release of drugs, biomolecules and other active ingredients for biomedical, pharmaceutical, food and consumer industries. Advances in microfluidic technologies have provided a new set of tools for unraveling the science behind formation of vesicles and fabricating novel vesicles. While traditional approaches for fabricating vesicles rely on self-assembly of amphiphiles, the precise control of flow afforded in microfluidic devices enables directed assembly of the amphiphiles. Thus, techniques such as hydrodynamic flow focusing, controlled emulsion-templating and pulsatile jetting offer unprecedented degree of control over vesicle structures. This creates new opportunities to engineer the structures of vesicles and tailor them for specific applications. In this review, we introduce current understanding behind different kinds of vesicles, survey conventional and microfluidic techniques for their formation, discuss new approaches of encapsulation and release of active ingredients in microfluidic vesicles, and point to future research and development in the area.

H. C. Shum (✉)
Department of Mechanical Engineering, University of Hong Kong, Pokfulam Road, Hong Kong, People's Republic of China
e-mail: ashum@hku.hk

J. Thiele
Radboud University Nijmegen, Institute for Molecules and Materials, Heyendaalseweg 135 6525 AJ Nijmegen, The Netherlands

S.-H. Kim
Department of Chemical and Biomolecular Engineering, Korea Advanced Institute of Science and Technology, Daejeon, South Korea

L. Wang (ed.), *Advances in Transport Phenomena 2011*,
Advances in Transport Phenomena 3, DOI: 10.1007/978-3-319-01793-8_1,

1 Introduction

1.1 What Are Vesicles?

Vesicles are compartments enclosed by a thin membrane, which is typically made up of bilayers of amphiphilic molecules, including phospholipids, skin lipids, block copolymers, and surfactants. Examples of these include liposomes [1], polymersomes [2, 3] and niosomes [4, 5]. Other vesicle-like structures include colloidosomes, whose membranes are not necessarily made up of bilayers of amphiphiles [6, 7]. Due to the co-existence of both hydrophilic, "water-loving" part and hydrophobic, "water-hating" part within the amphiphilic molecules, these molecules self-assemble in aqueous environments into a variety of aggregate structures, in which the hydrophobic parts avoid exposure to the aqueous surroundings. One such aggregate structure is vesicles. In the membranes of vesicles, two monolayers of amphiphilic molecules facing each other form a bilayer, as shown in Fig. 1a; if the membrane of the vesicle consists only of a bilayer, the vesicle is unilamellar (Fig. 1b). Vesicles with membranes made up of multiple bilayers stacked together are called multilamellar vesicles (Fig. 1c). The lamellarity of vesicles determines permeability of compounds through the vesicle membranes and also its ability to fuse with other vesicles. Vesicles are excellent structures for encapsulation of smaller compounds due to their ability to isolate the compounds in the core of the vesicles from the surroundings, providing a confined environment for processing and reaction of those compounds. Nature takes advantage of the self-assembly nature and encapsulation property to enable formation of biological cells, where important biochemical processes take place within phospholipid vesicles. Because of the resemblance of the vesicular structures to biological membranes, vesicles are also used as models in fundamental studies for understanding the physics and biology of natural membranes. Industrially, vesicles are also becoming important delivery vehicles for encapsulating drugs, flavors, colorings, vitamins and other important active ingredients for the pharmaceutical, food and beverage as well as cosmetics industries.

Vesicles also have the important attribute of fusing with one another; this allows components to be exchanged among biological cells for signaling and transport. Therefore, vesicles have the potential to be engineered to achieve targeted delivery of compounds to biological cells.

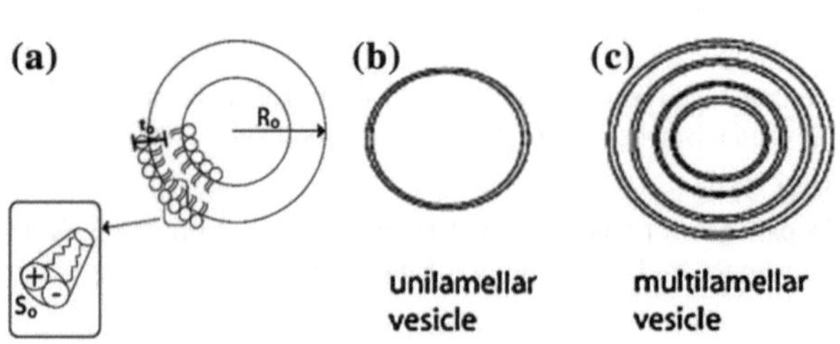

Fig. 1 (**a**) Structure of a vesicle with a bilayered membrane; Schematic of (**b**) a unilamellar vesicle, and (**c**) a multilamellar vesicle. Reproduced with permission from Ref. [8], © 2006 Elsevier

2 Vesicle Formation

Despite the vast number of amphiphiles and diverse application of vesicles, their formation mechanism is not yet understood in its entirety, and therefore remains an objective of many ongoing studies where different mechanisms have been proposed. For instance, polymersome formation is usually viewed as a two-step process [9]. Analogous to studies on liposomes by Lasic et al. [10] predicting a disk-like lipid micelle as an intermediate structure, copolymer molecules initially self-assemble into lamellar, sheet-like aggregates, that subsequently curve and close up to form vesicles, as shown in Fig. 2a. This process is driven by the energy loss owing to surface tension, which increases with the size of the disk-like bilayer, thus favoring spherical over flat bilayers [11, 12]. Recently, two alternative mechanisms have been proposed based on theoretical calculations.

In the first case, spherical micelles form from a homogeneous copolymer solution. After their rapid formation, they slowly coalesce and evolve into larger cylindrical or open micelles, which then curve into vesicular structures as shown in Fig. 2b [13–16]. Similar intermediates have been found in experiments using detergent depletion to elucidate the formation mechanism of phospholipid vesicles [10]. By steadily removing detergent from a solution of phospholipids dissolved in detergent micelles, the mixed micelles grow into aggregates by fusion from which lipid vesicles evolve.

In the second case, spherical micelles serve, again, as a starting point. They then grow by the uptake of further copolymer molecules into their interior in a condensation-evaporation process forming bilayered large micelles, so-called semi-vesicles, as shown in Fig. 2c [17]. Semi-vesicles are energetically unfavored and lower their energy by taking up solvent to give rise to vesicles [18].

Despite the number of possible mechanisms for polymersome formation, not all copolymers are able to self-assemble into vesicles; polymersomes only form under certain ranges of copolymers structures and compositions. The dimensionless packing parameter P dictates the molecular shape of copolymer molecules in solution, and thus the morphology of the corresponding copolymer aggregate upon phase separation of the hydrophobic and hydrophilic block. P is defined as the size of the hydrophobic block relative to the hydrophilic moiety [9].

$$P = \frac{v}{a \cdot l} \tag{1}$$

where v is the volume of the hydrophobic block, a the hydrophilic-hydrophobic interfacial area, and l the hydrophobic block length normal to the interface, as illustrated in Fig. 3. With increasing values of P, the morphology changes from spherical and globular structures, to toroidal, and to cylindrical aggregates, as shown in Table 1 [20, 21]. Whether vesicles can form or not also hinges upon the effective interaction parameter χ, which describes the strength of solvent-polymer interactions and is related to the dielectric constant ε and the solubility parameter δ [11, 22]. The polarity and dielectric constant of the copolymer solvent strongly

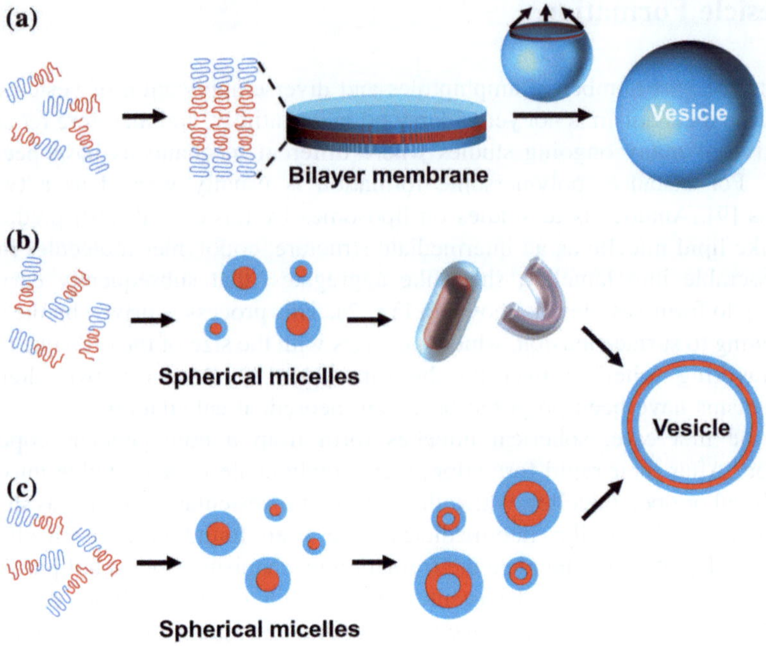

Fig. 2 Schematic of different polymersome formation mechanisms [9, 17]. A homogenous copolymer solution is assumed as a starting point. **a** Widely accepted two-step process involving the formation of a diblock copolymer bilayer followed by its closure and the formation of a hollow vesicle structure. **b–c** Proposed mechanisms of polymersome formation based on external potential dynamics simulations [10], molecular dynamics simulations [13] and density functional simulations [14] as well as dissipative particle [15] and Brownian dynamics studies [16], respectively. **b** Spherical micelles coalesce to interconnected worm-like and cylindrical micelles as well as open disc-like structures. Close-up of these structures give rise to the formation of vesicles, as experimentally observed [19]. **c** Spherical micelles grow by the uptake of copolymer molecules through a process analogous to evaporation and subsequent condension into bilayered micelles, so called semi-vesicles, which take solvent into their inside to reach the energetically more favorable vesicular shape. Reproduced with permission from Ref. [17], © 2009 Royal Society of Chemistry (RSC) Publishing

influences the repulsion between the hydrophilic blocks and, therefore, affects the morphology of the resultant copolymer aggregates including vesicles.

$$P = 1 - Hl + \frac{Kl^2}{3} \qquad (2)$$

with $H = \frac{1}{2}\left(\frac{1}{r_1} + \frac{1}{r_2}\right)$ and $K = \frac{1}{r_1 \cdot r_2}$.

In the case of a bilayer, $K = 0$, and $H = 0$. Insertion into (1–2) gives 1, as shown in Table 1.

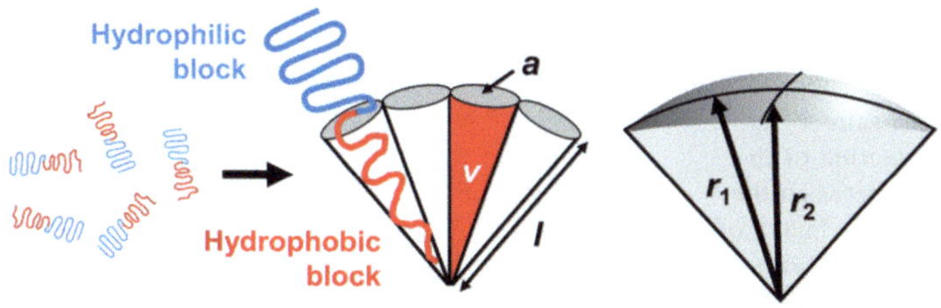

Fig. 3 Illustration of the packing parameter P in terms of the interfacial area a, the hydrophobic volume of the copolymer v and the chain length normal to the interface l (*left*), as well as its relation to the interfacial mean curvature and Gaussian curvature, described by the curvature radii r_1 and r_2 (*right*). Adapted with permission from Ref. [9], © 2003 WILEY-VCH

Table 1 Packing parameter P of different aggregates and their corresponding mean curvature H and Gaussian curvature K, expressed in terms of the two radii of curvature r_1 and r_2

Shape	P	r_1	r_2	H	K
Sphere	$\frac{1}{3}$	r	r	$\frac{1}{r}$	$\frac{1}{r^2}$
Cylinder	$\frac{1}{2}$	r	∞	$\frac{1}{2r}$	0
Bilayer	1	∞	∞	0	0

Typical range of P for different aggregate shapes is as follows: Sphere ($P \leq \leq 1/3$), cylinder ($1/3 < P \leq 1/2$), and bilayer ($P > 1/2$). As the vesicle shape is mainly determined by interfacial curvature, the packing parameter is closely related to the mean curvature H, and Gaussian curvature K of the interfacial surface through the two radii of curvature r_1 and r_2 [21, 23]. Reproduced with permission from Ref. [9], © 2003 WILEY-VCH

The size of the hydrophobic block, which dictates the bilayer thickness of the polymersome and thus the elasticity and stability of the membrane, provides a simple scaling of the copolymer membrane thickness d,

$$d \sim (M_h)^b \tag{3}$$

where b is a parameter describing the folding state of the polymer chain with $b = 1$ for a fully stretched polymer chain, $b = 0.5$ for an ideal random coil, and $b \approx 0.55$ in a polymersome, and M_h the mean molecular weight of the hydrophobic block [20, 24, 25]. M_h can be estimated from the number average molecular weight M_N and the hydrophilic fraction f, which often ranges from 0.3 to 0.4 [26].

$$M_h \sim M_N(1 - f) \tag{4}$$

As the amphiphiles on the inner and outer surfaces of the polymersome bilayer are trapped in a non-equilibrium state in the fabrication process, the bilayer spontaneously curves to minimize the bending energy due to the different number of amphiphiles on the inner and the outer copolymer monolayer [9, 11]. This, in turn, allows for tailoring of vesicle size and morphology by the preparation method and the experimental conditions, resulting in a diverse ensemble of polymersomes, where each geometry represents a state of minimal bending energy. Applying the

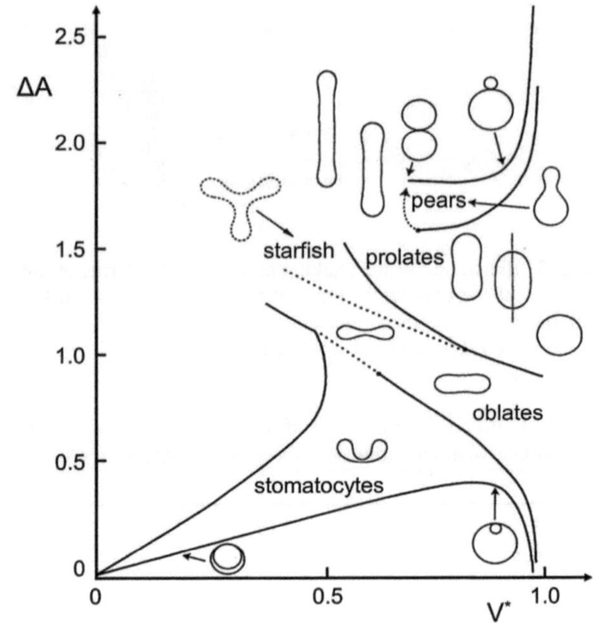

Fig. 4 Phase diagram of theoretical polymersome shapes. The dimensionless volume-to-area ratio V^* is plotted as a function of the area difference between the inner and the outer layers of the bilayer. Adapted with permission from Ref. [27, 28] © 1995 Cell Press

area difference between the inner and outer bilayer surface, A_{in} and A_{out}, respectively,

$$\Delta A = A_{in} - A_{out} \tag{5}$$

and the volume-to-area ratio $V*$

$$V^* = \frac{V}{\frac{4}{3}\pi R_A^3} \tag{6}$$

with $R_A = \left(\frac{A}{4\pi}\right)^{\frac{1}{2}}$ and the vesicle volume V, where $V^* = 1$ for spherical vesicles, the different vesicle shapes can be mapped in a phase diagram, as shown in Fig. 4.

Depending on the number of bilayers that are interlaced with one another, vesicles can be broadly distinguished between unilamellar, oligolamellar and multilamellar structures, as shown in Fig. 5 [29]. Unilamellar vesicles are further classified as small, large or giant vesicles; vesicles encapsulated within vesicles are defined as multivesicular vesicles.

3 Conventional Techniques for Fabrication of Vesicles

Traditionally, vesicles are fabricated via the self-assembly of the constituent amphiphilic molecules. This approach for generating vesicles is applied in two common vesicle fabrication techniques, namely, hydration and electroformation.

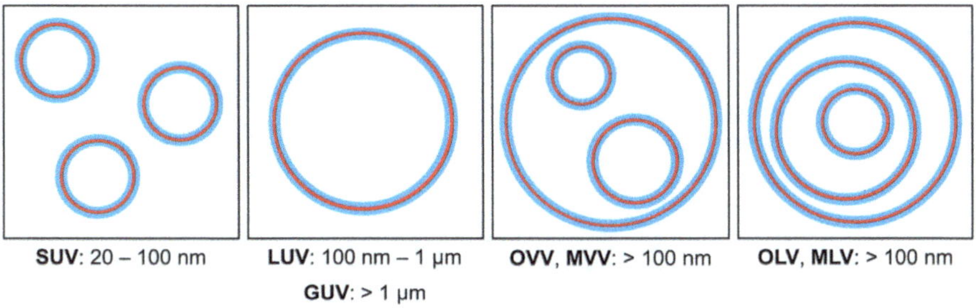

Fig. 5 Classification of vesicle structures. Depending on the number of nested bilayers, vesicles can be classified as unilamellar, oligolamellar and multilamellar. Unilamellar vesicles have only one bilayer, Depending on the vesicle size, unilamellar vesicles can be classified as small unilamellar vesicles (*SUV*), large unilamellar vesicles (*LUV*) and giant unilamellar vesicles (*GUV*). Oligolamellar vesicles (*OLVs*) and multilamellar vesicles (*MLVs*) have membranes making up of several and multiple bilayers respectively. Vesicles that are encapsulated within vesicles are specified as oligovesicular vesicles (*OVV*) and multivesicular vesicles (*MVV*)

In both methods, amphiphilic molecules are first dissolved in an organic solvent, such as chloroform, where the molecules are dispersed. The organic solvent containing the amphiphilic molecules is then left to evaporate on a solid substrate, such as glass slides, silicon wafers, and indium tin oxide (ITO) slides. As the solvent is gradually removed through evaporation, the amphiphilic molecules form a dry film on the surface of the substrates. Subsequently, a drop of the solution containing the compounds to be encapsulated is placed on the dry film and gently stirred. During this hydration process, the dry film of amphiphilic molecules is soaked and folds around the solution, achieving the necessary encapsulation. However, with the hydration process, the structure of the membrane is not well controlled and the resultant vesicles are often multilamellar. To improve the uniformity of the vesicles and prepare unilamellar vesicles, an alternating electric field is applied to perturb the film of amphiphilic molecules in a process called electroformation [30]. Despite the improved uniformity in the lamellarity, the resultant vesicles remains polydisperse in size and the encapsulation efficiency is typically limited to only 35 % or lower [31]. Various attempts have been made to modify these conventional techniques to improve the size distribution of the vesicles and encapsulation efficiency, for instance, through electroformation or dewetting from micropatterned substrates [32, 33], as shown in Figs. 6 and 7. This approach takes advantage of the ability to precisely pattern substrate via lithographic techniques such as photolithography to achieve 'top-down' control of micrometre-sized features, such as vesicle diameter. Moreover, vesicle sizes can also be further modified by swelling [34] and extrusion [35, 36] techniques. By extruding larger vesicles through membranes with pores, the size of the vesicles can be reduced and the size uniformity can be improved [37].

Another approach to fabricate vesicles is through emulsion-templating. In this approach, a water-in-oil (W–O) emulsion is stabilized by the amphiphilic molecules for the inner leaflet. By passing the resultant water droplets through a second

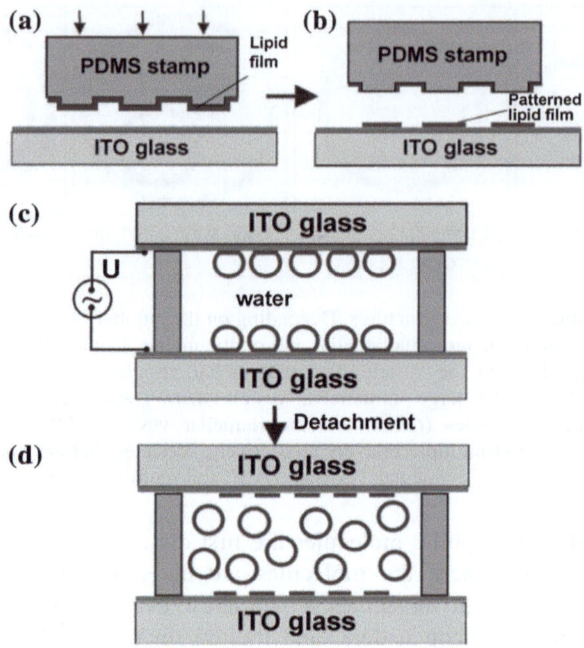

Fig. 6 Preparation of monodisperse giant liposomes by microcontact printing technology: (**a**) and (**b**) the lipid (stained with a fluorescent dye) is deposited on an ITO glass by using a PDMS stamp of square grid pattern. **c** The liposomes are grown on the ITO slide in water by electro formation in AC field. **d** The monodisperse liposomes are released from the ITO slide after application of ultrasound (ultrasonic bath) or low frequency AC field. Reproduced with permission from Ref. [33], © 2003 Royal Society of Chemistry (RSC) Publishing

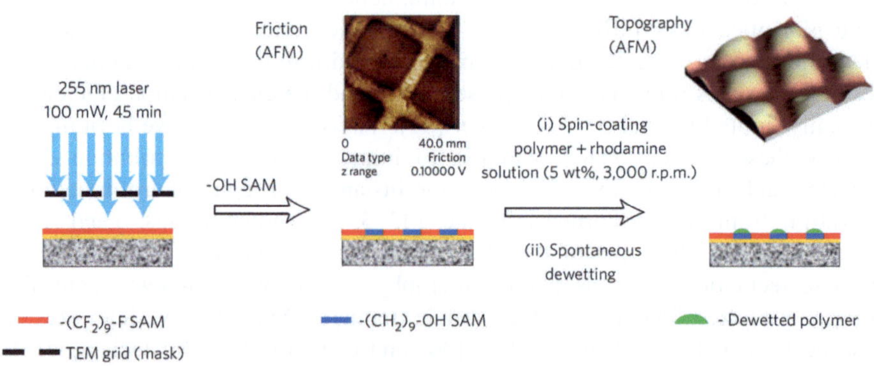

Fig. 7 Schematic representations of polymersome formation through dewetting on micropatterned substrates. Procedure for the formation of patterned hydrophilic, fluorocarbon-decorated self-assembled monolayers (*SAMs*) and the spontaneous dewetting of a hydrophobic–hydrophilic block copolymer resulting in micrometre-sized domains of polymer. Friction atomic force microscopy (*AFM*) was used to characterize the fluorocarbon (*light*) and hydrophilic (*dark*) domains. Reproduced with permission from Ref. [32], © 2009 Nature Publishing Group

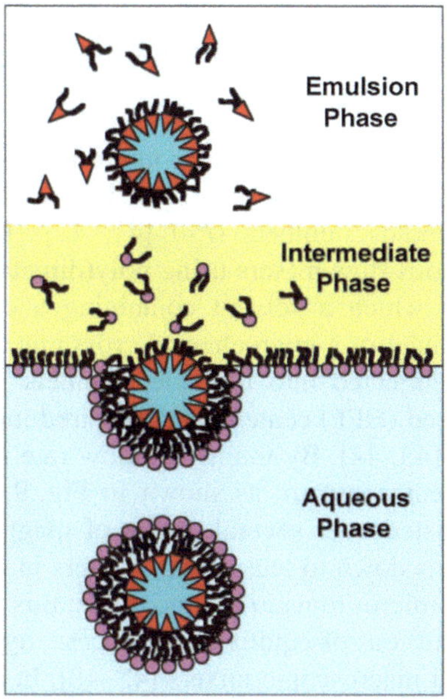

Fig. 8 Schematic illustration of the emulsion-templating technique used to engineer asymmetric vesicles. The sample is composed of three parts: an inverted emulsion where water droplets in lipid-saturated oil are stabilized by lipid molecules destined for the *inner* leaflet; an intermediate phase of lipid-saturated oil heavier than the inverted emulsion phase, and whose lipids form a monolayer at the oil water interface; and the *bottom* aqueous phase, which receives the final asymmetric vesicles. The lipids in the intermediate phase are completely different from those in the inverted emulsion and form the *outer* leaflet of the bilayer; the final structure is an asymmetric vesicle. Reproduced with permission from Ref. [39], © 2003 National Academy of Sciences of the United States of America

oil–water interface, the droplets are coated with another monolayer of amphiphilic molecules and vesicles are produced [38]. A schematic of the technique is illustrated in Fig. 8. This approach offers greater tenability in the composition of the two leaflets of the bilayer and enables the fabrication of asymmetric vesicles, which are vesicles with different compositions in the two leaflets of the bilayer membrane [39].

4 Microfluidic Approaches for Fabrication of Vesicles

Advances in microfluidic technologies have led to more precise fluidic control, creating new opportunities for preparation of vesicles [40]. By confining the size of the channels in which fluids flow through, stable lamellar flow can be achieved; this eliminates turbulent mixing as a potential source of size polydispersity in vesicle fabrication. With the variety of available techniques for fabrication of microfluidic devices, the geometry of the devices can be easily customized, making it possible to explore different device design for fabrication of vesicles.

- Hydrodynamic Flow-focusing

The most obvious advantage provided by microfluidic devices is the excellent control over fluid volume on the microscale, thereby providing laminar, stable flow conditions at low Reynolds numbers (*Re*) [41, 42]. This has stimulated the development of continuous-flow mixers using poly(dimethyl siloxane)- or PDMS-based microfluidics, in which a solvent containing a compound of interest is injected into the center inlet of a microchannel cross junction and narrowed into a jet by solvent streams injected into the side channels. At the interface of the hydrodynamically focused (HFF) center jet, the desired mixing process is initiated by molecular diffusion [43, 44]. By tuning the flow rate ratio (FRR) between the side streams and the center stream, as shown in Fig. 9, the width of the flow-focused jet can be adjusted over several orders of magnitude, allowing the formation of stable fluid jets down to tens of nanometers in diameter, and controlled mixing timescale in the micro- to nano-second range; this greatly facilitates access to reaction kinetics and out-of-equilibrium process dynamics that cannot be resolved in conventional macroscopic mixers [45–48]. In addition, as microfluidic HFF naturally requires small sample volumes, valuable biomacromolecules such

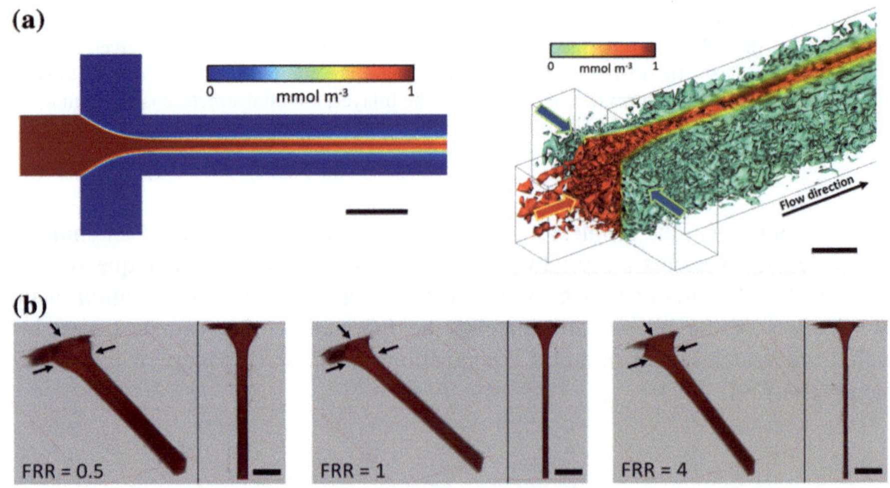

Fig. 9 Vesicle formation by hydrodynamic flow focusing based on simulation and experiment. **a** Fluid flow simulation of diffusion-based mixing at the interface of a flow-focused fluid stream using COMSOL Multiphysics. The concentration profile is rendered by applying 170,000 finite elements for the 2D model (*left*) and 62,000 finite elements for the 3D model (*right*). The scale bars denote 20 μm. **b** 3D reconstructions of stacks of 2D CLSM images: an aqueous solution of Rhodamine B is flow-focused by ethanol in a microfluidic cross junction. Control over the width of the flow-focused stream is exerted by tuning the flow rate ratio (*FRR*) which is the ratio of the volumetric flow rate in the side and center channel. The scale bar denotes 100 μm

as proteins or DNA can be studied under mechanical stress utilizing the influence of geometric constraints on the flow at the center jet's interfaces [48, 49]. As the time evolution of the reaction scales with length from the initial mixing location in the outlet channel of the microfluidic mixer due to low Reynolds flow, the concentration inside the flow-focused fluid jet can be precisely determined at any point of the reaction.

However, in planar microfluidic devices, observation of the two dimensional HFF patterns are limited to a small area within the X, Y plane, as shown in Fig. 9. Therefore, most information on the velocity profile and distribution of reagents inside the flow-focused jet is out-of-plane and remains undetected. These limitations are overcome by shaping flow-focused jets into a 2D plane using multilayered PDMS-based microfluidic devices. Tight planar fluid jets with thicknesses of two to three microns are formed using multilayered PDMS-based microfluidic devices. This approach has been extended to three-dimensional flow-focused jets, thereby reducing sample consumption and improving velocity field distribution within the jet [45].

The inherent properties and outstanding performance of microfluidic HFF have inspired a wide range of applications in the field of complex micro- and nanostructure fabrication by controlled self-assembly of the respective building blocks at the liquid–liquid interface of flow-focused jets [50, 51]. This includes the formation of multi-component nanoparticles as well as liposomes and polymeric vesicles that can be engineered in sizes, monodispersity and complexity that have not been achieved using conventional bulk and microfluidic techniques, as exemplarily shown in Fig. 10. In the following section, selected applications of microfluidic HFF for fabricating vesicles, which are the focus of this review, will be discussed.

Kinetically trapped, nanosized liposomes from dimyristoylphosphatidylcholine (DMPC) and cholesterol have been formed in microfluidic cross junctions via controlled self-assembly [52]. Liposome size is controlled by FRR and thus the width of the lipid-loaded center stream and, to a less extent [53], the shear stress applied to the lipids at the liquid–liquid interface. In a simple nucleation/growth model that is applicable to both liposome and polymersome formation, liposome nuclei form at the interface of a flow-focused center stream and grow by uptake of lipids. Since the amount of lipids is proportional to the width of the focused stream in a given volume element, larger liposomes self-assemble from focused jets with larger width and, vice versa, smaller liposomes from focused streams with smaller width. Recent studies have suggested that liposome size distribution further depends on the total flow rate and the size of the microfluidic device [51]. The variety of control mechanisms over vesicular self-assembly allows one to influence vesicle polydispersity compared to traditional bulk methods, which have been described earlier, and to avoid post-processing steps to manipulate shell characteristics and size. Thereby, HFF can also be applied for forming vesicles with tailored size in the range of 50–150 nm, which is the optimal size for biomedical applications including tumor-targeted drug delivery, preserving of cell viability. Such vesicles are apt for benefiting from the enhanced permeability and retention

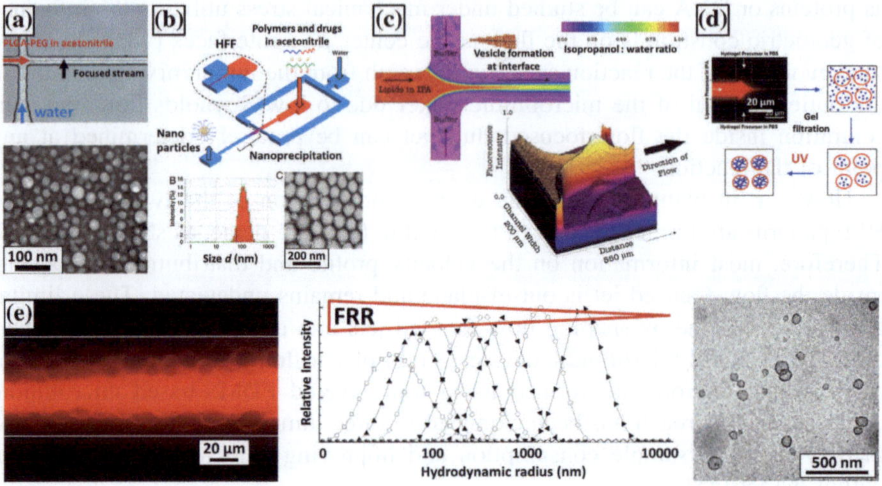

Fig. 10 Application of hydrodynamic flow-focusing for the fabrication of complex nanoparticles and vesicles in microfluidics. **a** Nanoprecipitation of drug-loaded poly(lacticco-glycolic acid)-*b*-poly(ethylene glycol) (PLGA-PEG) particles. Reproduced with permission from Ref. [60], © 2008 American Chemical Society. **b** Fabrication of poly(lactic acid)-platin(IV)prodrug / docetaxel nanoparticles for combined drug therapy. Reproduced with permission from Ref. [61], © 2010 National Academy of Science of the United States of America. **c** Size-controlled formation of liposomes loaded with the fluorescent membrane dye $DiIC_{18}$. Reproduced with permission from Ref. [52], not subject to U.S. Copyright. Published 2004 American Chemical Society. **d** Preparation of liposome-poly(*N*isopropylacrylamide) hybrid nanoparticles with temperature-switchable vesicle size. Reproduced with permission from Ref. [56], © 2010 American Chemical Society. **e** Fabrication of poly(2-vinylpyridine)- *b*-poly(ethylene glycol) (P2VP-PEG) polymersomes. Ultra-giant vesicles form at the interface of the flow-focused copolymer stream loaded with Rhodamine B and adhere to the microchannel surface (*left*). The vesicle size is tuned over several orders of magnitude by controlling the flow rate ratio (*FRR*) of the copolymer-loaded center stream and the aqueous side streams, as determined by dynamic light scattering (*middle*). Smaller vesicles are characterized by cryo-TEM (*right*). Reproduced with permission from Ref. [58], © 2010 American Chemical Society

effect (EPR) [54, 55]. The ability to load vesicles in situ with actives and dyes using HFF has stimulated the development of highly sophisticated multi-component vesicles including liposome-hydrogel hybrid particles, so-called lipobeads, as shown in Fig. 10d [56]. The phospholipid vesicles encapsulate a solution of photopolymerizable *N*-isopropylacrylamide, which forms a thermo-responsive polymer core and increases the mechanical stability of the liposome shell. In addition, control over vesicle size is extended beyond FRR by utilizing the swelling/shrinkage of pNIPAAm around its lower critical solution temperature (LCST) so that the size of the liposome shell becomes highly sensitive to temperature changes [57].

The initial approach to form lipid vesicles by HFF has also been adapted to control the self-assembly of amphiphilic diblock copolymers into polymersomes, as recently reported by Thiele et al. [58]. In contrast to earlier studies on lipid vesicles, HFF enables the tuning of the size of polymeric vesicles over several orders of magnitude from less than 40 nm to more than 2 μm in the same experimental device, as summarized in Fig. 10f. However, the use of organic solvent in the production of polymersomes hampers potential biomedical applications. This limitation can be overcome by introducing pH-sensitive diblock copolymers such as PMPC-*b*-PDPA, which self-assemble into polymersomes from acidic aqueous solutions that are flow-focused by basic buffer solutions, creating a pH gradient in the outlet channel of the microfluidic device [59]. With this approach, the use of organic solvent can be avoided.

Apart from hydrodynamic flow-focusing, microfluidic technologies have enabled generation of highly controlled single and multiple emulsions, which are excellent templates for formation of vesicles.

- Emulsion-templating with Single Emulsions

Using microfluidic technologies, monodisperse W–O emulsions of different sizes can be routinely generated. This approach of droplet generation can be applied for the fabrication of emulsion-templated vesicles, as shown in Fig. 11. Monodisperse water droplets stabilized by phospholipids dissolved in an organic solvent are first generated in a microfluidic device. The droplets are subsequently coated with another layer of phospholipids by another liquid–liquid interface laden

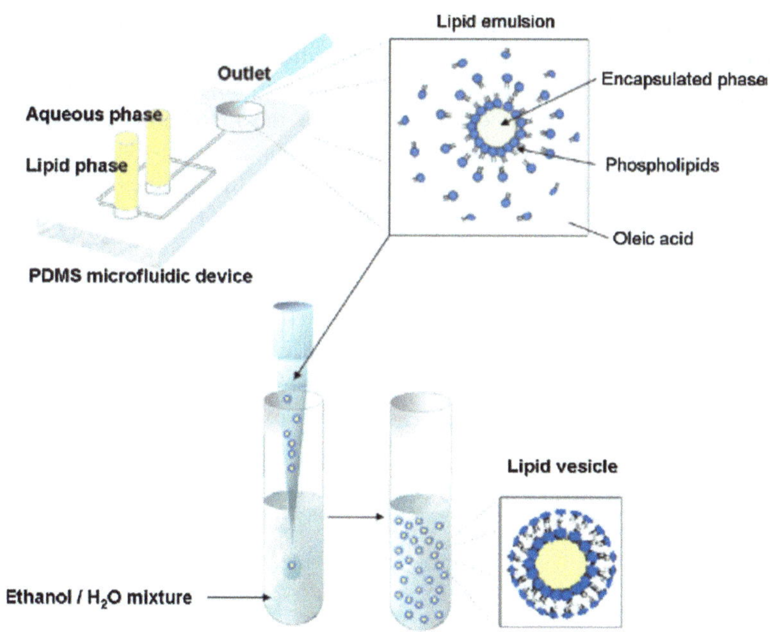

Fig. 11 Preparation of single-emulsion-templated phospholipid vesicles. Reproduced with permission from Ref. [62], © 2006 American Chemical Society

with phospholipids [62]. Using this approach, monodisperse phospholipid vesicles can be generated with high encapsulation efficiency.

- Emulsion-templating with Double Emulsions

Apart from single emulsion templates, vesicles can also be fabricated using double emulsion droplets as templates. Water-in-oil-in-water (W–O–W) double emulsions can be routinely generated with excellent size uniformity and 100 % encapsulation efficiency using microfluidics. Double emulsions and vesicles share similar core–shell structures; therefore, double emulsion droplets are ideal templates for vesicles. A schematic of the double-emulsion-templated approach for forming vesicles is shown in Fig. 12. To convert double emulsion droplets into vesicles, the two interfaces need to be first adsorbed by the amphiphilic molecules; therefore, the amphiphilic molecules should contain one block that has an affinity for the inner or the outer aqueous phases, and another block that has an affinity for the middle organic solvent phase. The simple structures that satisfy this requirement are diblock copolymers and phospholipids that have hydrophilic and hydrophobic groups. Before the double emulsion drops form vesicles, the solvents in the middle phase need to be removed, for instance, through solvent evaporation [63–65]. Nevertheless, the double emulsion drops should maintain their stability during the solvent removal step. Recently, it has been shown that by using a mixture of two organic solvents in the middle phase, where the amphiphilic molecules are soluble in one volatile organic solvent but insoluble in the other less volatile organic solvent, the two surfactant-laden interfaces, namely the inner-middle, and the middle-outer interfaces, attract towards each other forming the membrane of the resultant vesicles [66]. Due to the attraction, the double emulsion drops undergo a dewetting transition, where the middle phase dewets and get collected into a single droplet on the surface of the inner droplet of the double emulsions. As a result, the double emulsion adopts an acorn-like geometry. The remaining middle phase that forms a drop attached to the vesicles can be removed subsequently due to flows inside the microfluidic device. This approach represents a promising strategy for fabricating monodisperse vesicles with almost perfect encapsulation efficiency.

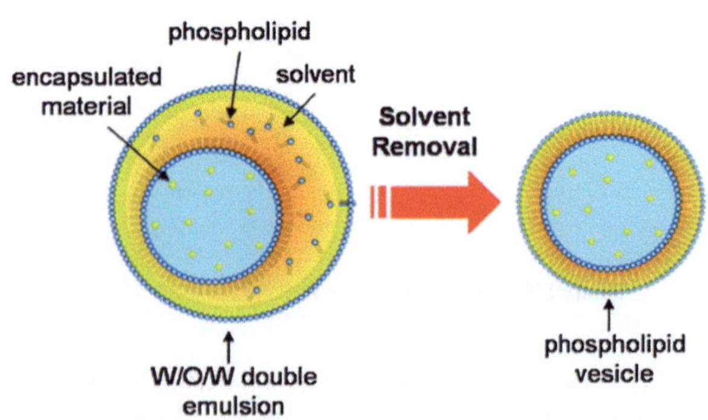

Fig. 12 Preparation of phospholipid vesicles using double emulsion as templates. Reproduced with permission from Ref. [64], © 2008 American Chemical Society

Other Microfluidic Approaches

Apart from emulsion-templating methods, various microfluidic approaches have been studied to make monodisperse vesicles. One way to produce monodisperse giant vesicles is microfluidic jetting of aqueous solution against a planar bilayer [67, 68]. The high inertia of a pulsed jet flow makes protrusion through local deformation of the planar bilayer and induces a breakup of the bilayer, resulting in a vesicle and intact planar bilayer.

In a similar manner with double-emulsion-templating methods, a planar bilayer can be prepared by overlapping of two interfaces between water and oil which have a dense array of amphiphiles. To prepare a wide planar bilayer, a hydrophobic double-well structure has been employed, which is composed of two circular cells connected with a waist of approximately 2 × 2 mm in a dimension, as shown in Fig. 13a. Into the waist of the double-well, an oil solution containing the amphiphiles is introduced and subsequently, two aqueous solutions are added into each circular cell. The oil solution wets the wall due to the hydrophobic nature of the surface, and forms a connecting thin film in the waist which is surrounded by two water drops. Two interfaces of the film of oil solution are stabilized through adsorption of the amphiphiles from the oil. Although the liquid film is relatively thick in the proximity of the solid wall of the waist, it becomes very thin far away from the walls. Therefore, the relative contents of oil become very small at the center of the film, resulting in a planar bilayer of the amphiphiles.

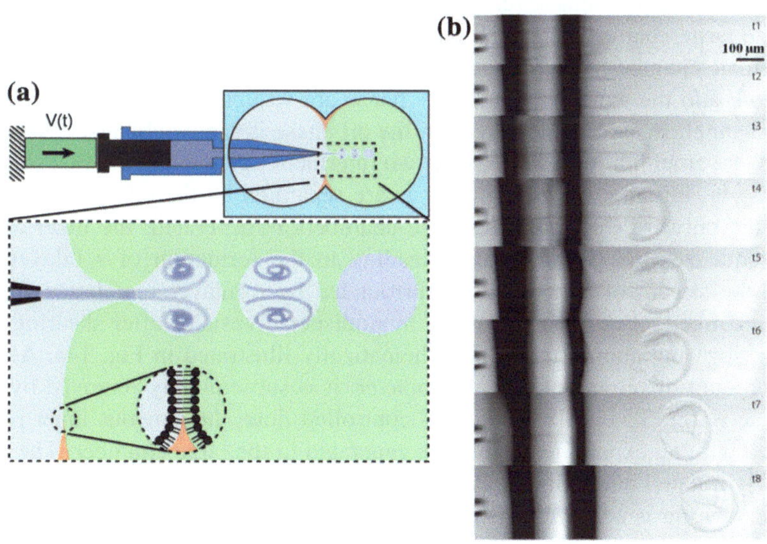

Fig. 13 **a** Schematics of microfluidic jetting for generation of vesicles. **b** Series of still images showing formation of vesicles from planar bilayer. Reproduced with permission from Ref. [68], © 2008 National Academy of Sciences of the United States of America

When aqueous solution flows out in a pulse from the nozzle into stationary fluids, shear stress causes formation of a vortex ring. The energy of the vortex is sufficient to deform the bilayer membrane, producing protrusion in the bilayer membrane. The protrusion grows due to the momentum of the aqueous stream; this causes the bilayer to curl and fold into a spherical shape connected to the original bilayer through a long and narrow neck. The aqueous stream which fills the protrusion is dominantly the aqueous solution injected from the nozzle but, contains the stationary fluid at a volume fraction of approximately 25 % of the stream. When the inertia is enough to induce pinch-off of the neck, the spherical vesicle separates from the original bilayer and closes a hole to reduce exposed area of hydrophobic part of the amphiphiles to water. In addition, the deformed and ruptured bilayer at the waist is self-repairing and achieves the initial planar shape by retraction and supplement of the amphiphiles from thick oil film. Therefore, sequential generation of monodisperse vesicles can be achieved by repeated pulsed jetting.

The size of the resultant vesicles depends on the pulse volume and velocity. Under normal conditions, vesicles fabricated with this approach have a diameter of about 5–6 times that of a nozzle. Low pulse volume and high velocity can reduce the diameter of the vesicles to about 1–2 times that of a nozzle. However, since smaller nozzle necessitates larger pressure drop, it is difficult to achieve a strong jet in a small nozzle. Therefore, the minimum size of vesicles achievable in this approach with practical operating conditions is approximately 100 μm.

To produce cell-sized vesicles, the planar bilayer membrane is prepared in a microfluidic device with an alternative design shown in Fig. 14 [69]. The main channel of the device has several small chambers on the wall, which are connected to some large chambers through narrow channels. To form bilayer membrane in each small chamber, the aqueous inner phase, the middle phase of amphiphile-laden oil, and the aqueous continuous phase are sequentially injected through the main channel. In each small chamber, the oil phase is sandwiched by two aqueous phases and amphiphiles adsorb at the two interfaces as shown in the first schematic of Fig. 14a. Initially, the two interfaces are separated by the oil phase. However, when the outward flux from the large chambers is generated, the inner aqueous phase squeezes the two interfaces, leading to the formation of a bilayer. As a result, the stream of aqueous core surrounded by a bilayer membrane from the small chamber breaks and folds into monodisperse vesicles after shearing off by the flowing continuous stream, as schematically illustrated in Fig. 14a. Although trace amount of oil can exist in the bilayer, it is subsequently removed by dissolution in water. To ensure gentle and controlled flow, the aqueous inner phase is pumped with an expanding air bubble generated in the large chamber. The bubble is slowly inflated due to localized heating of an aluminum pattern surrounding the bubble; heating is achieved by irradiation with an IR laser, as shown in Fig. 14b, instead of a pulsed jetting. The breakup of the core–shell stream is shown in Fig. 14c. The mechanism by which a breakup takes place is similar with drop breakup in T-junction [49]. Therefore, size of vesicles depends on the flow rate of the continuous phase and width of the small chamber, regardless of the flow rate of

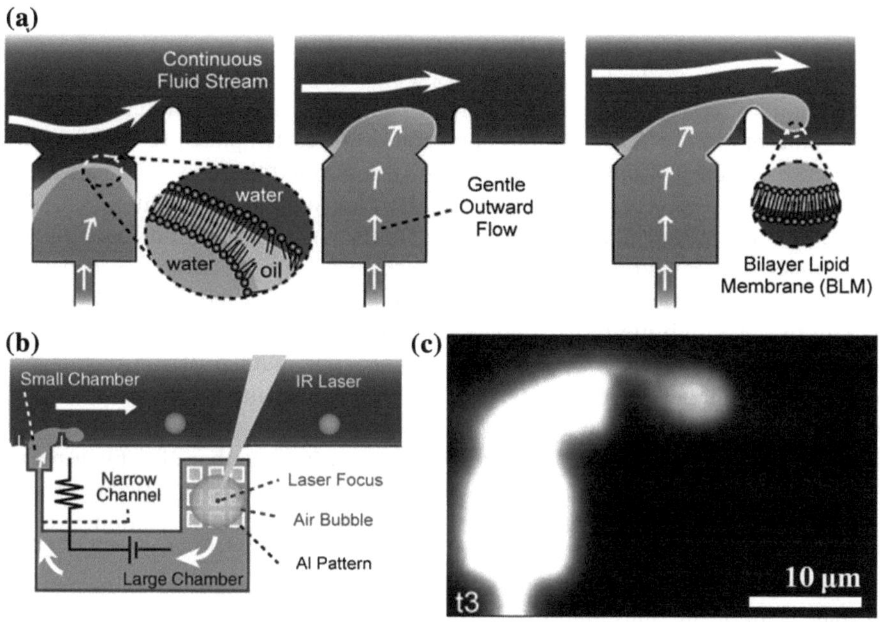

Fig. 14 **a** Schematic of a bilayer breakup for vesicle formation. **b** Schematic of IR-induced bubble generation for an gentle outward flux. **c** Fluorescence microscope image showing generation of vesicle through a breakup of aqueous core and bilayer shell flow; inner water phase contains fluorescent calcein. Reproduced with permission from Ref. [69], © 2009 WILEY-VCH

the outward flux of the aqueous inner stream. This method can produce monodisperse vesicles with cell-sized diameters of 10 μm with high encapsulation efficiency.

5 Fabrication of Novel Vesicles Using Microfluidics: Multi-Compartment Vesicles and Vesicles-in-Vesicles

Structural diversity of vesicles is very important to achieve programmed release of multiple distinct ingredients. For example, vesicles with multiple compartments are useful for a simultaneous release of multiple ingredients which are separately encapsulated and stored without cross-contamination between incompatible ingredients. However, conventional approaches cannot incorporate multiple compartments into a single vesicle without post-processing, because self-assembly of amphiphilic molecules without any guiding templates always results in spherical vesicle with a single compartment. Although controlled clustering of several vesicles can be employed to incorporate multiple compartments into a single vesicle, resultant vesicles are highly polydisperse in size and structure due to a broad size distribution of building block vesicles; [70] to induce adhesion and bonding between the vesicles, bilayers in the vesicles have to be modified by complementary single stranded DNA.

By contrast, microfluidic approach based on double-emulsion drops enables the production of vesicles with multicompartment in a simple and facile manner. Using microfluidic W/O/W double emulsion drops as templates for the vesicles, the number of inner droplets can be precisely controlled through the flow rates of the phases involved, as shown in Fig. 15a. By choosing an amphiphile-containing oil phase as the middle phase of the double emulsions, adhesion of the amphiphile-laden inner-middle and middle-outer interfaces can be induced; thus vesicles can be formed through dewetting and subsequent separation of the dewetted oil drop. In addition, the multiple inner droplets also adhere strongly with each other; this results in multicompartment polymersomes [71]. The number of compartments is fixed by the number of inner water drops. During dewetting, inner water drops adhere with neighbors due to attraction, which do not allow significant rearrangement. Therefore, resultant vesicles have a constant number of compartments, but different spatial arrangement of the compartments. The multicompartment vesicles consisting of eight cores are shown in Fig. 15b, which exhibit two different configurations.

The modification of microfluidic devices enables the encapsulation of multiple distinct ingredients into multiple compartments. For example, by employing injection capillary with two bores, double-emulsion drops containing two different inner water drops can be prepared, which can be converted to dumbbell-shaped vesicles encapsulating two different species without mixing, as shown in Fig. 15c.

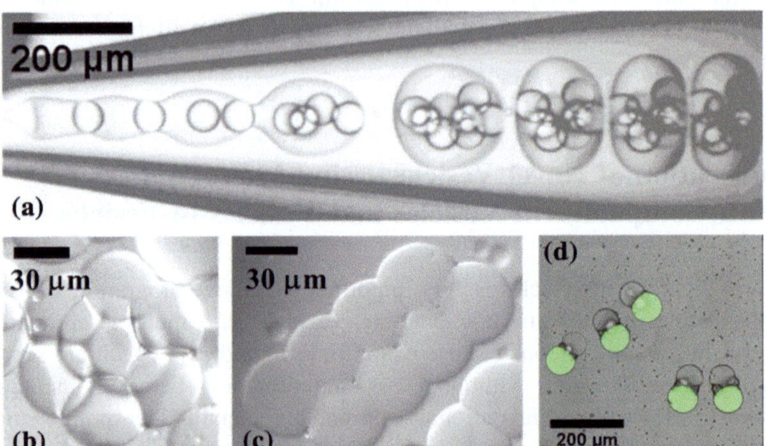

Fig. 15 **a** Optical microscope image showing generation of double-emulsion drops containing eight inner drops. **b** Optical microscope images showing nonspherical vesicles with eight compartments in two different configurations. **c** Confocal microscope image of dumbbell-like vesicles whose two cores encapsulate two distinct components. Reproduced with permission from Ref. [71], © 2011 WILEY-VCH

In this way, the number of encapsulants can be further increased by using injection capillary with multi-bores for inner drops that contain different species.

Another type of useful vesicle structure is multiple vesicles or multivesicular vesicles, which is vesicles-in-vesicle structure. Unlike multicompartment vesicles which have several compartments in the same level, the multiple vesicles have hierarchical compartments in multiple levels. Therefore, multiple vesicles enable the sequential release of multiple distinct ingredients from the outermost compartment, while avoiding cross-contamination during encapsulation, storage, and release. Double vesicle structures, or vesosomes, have been prepared by two-step bulk process to achieve sequential and extended release or long-term storage of multiple components [72]. However, they are randomly structured and do not allow systematic incorporation of different components.

Microfluidic approach enables controlled preparation of multiple vesicles. Using capillary microfluidic devices, monodisperse vesicles are prepared from templates of double-emulsion drops, which in turn are encapsulated in the innermost drops of a second level of double-emulsion drops using another microfluidic device, as shown in Fig. 16a. Therefore, small vesicles are confined in outer vesicles through formation of outer bilayer membrane from the double-emulsion drops, resulting in double vesicles [73]. During the encapsulation of the vesicles in

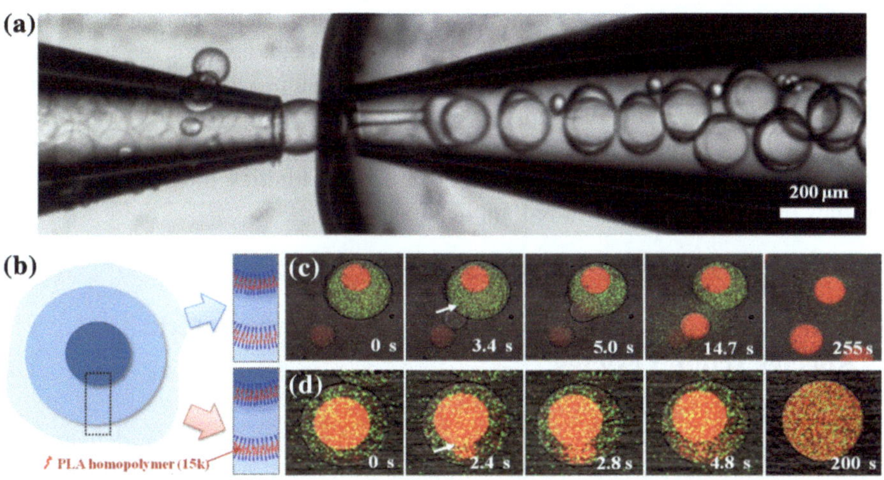

Fig. 16 **a** Optical microscope image showing encapsulation of monodisperse vesicles into the innermost drops of double-emulsion drops. **b** Schematic illustration of double vesicles. **c** A series of confocal microscope images showing selective dissociation of the outer membrane of double vesicles, consisting of amphiphilic diblock-copolymer bilayer for both the inner and outer membranes; the double vesicle is dispersed in a mixture of water and ethanol to induce a rupture. **d** A series of confocal images showing selective dissociation of the inner membrane of double vesicle, consisting of amphiphilic diblock-copolymer bilayer for the inner membrane and a hydrophobic homopolymer-loaded bilayer for the outer membrane, in the same mixture of water and ethanol. Reproduced with permission from Ref. [73], © 2011 American Chemical Society

the second step of double emulsification, the vesicles trigger breakup of the jet of the middle oil phase; in this manner, each resultant double emulsion drops only contain one vesicle. This encapsulation process can be repeated in a third step of double emulsification to form triple vesicles consisting of a vesicle-in-vesicle-in-vesicle structure, further enhancing the flexibility of the subsequent controlled release profile. These multiple vesicle structures show sequential and extended releases of multiple species encapsulated when they are exposed to specific chemical environment that induces dissociation or degradation of the bilayer membranes. In addition, modification of bilayer stability enables the programmed release of multiple active ingredients encapsulated. For instance, if the outer membrane of the double vesicles bursts before the inner membranes does, species encapsulated in the outer compartment will be released before those encapsulated in the inner vesicular compartments. This enables sequential release of different active ingredients. However, if the inner membrane is destabilized before the outer membrane is, internal mixing of ingredients within the intact outer membrane will take place before subsequent release to the surrounding environments following subsequent breakage of the outer membranes. This has great potential for enabling reaction of the active ingredients before release. These programmed releases are shown in Fig. 16c, d.

The combination of the two different structures, namely, multiple compartments in the same level and hierarchical compartments in different levels, will enables programmed release of multiple distinct components in desired patterns. The stepwise release in time of multiple components can be achieved by multiple vesicles structure, where time gap between release steps can be tuned by manipulating the stability of the bilayers. In addition, simultaneous release without cross-contamination during storage can be achieved using multicompartment vesicles. These novel structures create unprecedented opportunities to tailor the release profiles of actives for the desired applications.

6 Triggered and Sustained Release of Encapsulated Species from Vesicles

Polymersomes have become a versatile tool for drug encapsulation, delivery and release, as they offer a large variety of mechanisms to disassemble in response to specific external stimuli in a controlled manner. Polymeric building blocks have been developed which can be degraded upon changes of temperature, pH as well as by oxidation/reduction and enzymes [74].

The membranes of vesicles are typically permeable to water; this is an important feature that enables complex biological processes to take place inside biological cells. This also provides a strategy to release active ingredients encapsulated inside vesicles. The direction at which water moves across vesicle membranes is typically governed by the osmotic pressure difference between the inside and the outside environment of the vesicles. The osmotic pressure, $\pi_{osm} = cRT$, is related to the concentration of solutes, where c is the molar concentration of the solutes, R is the gas constant and T is the temperature. When the osmolarity, which refers to the molarity of osmotic relevant molecules that cannot diffuse across membranes, is higher in the outside environment than in the core of the vesicles, water tends to get sucked out of the vesicles; when the osmolarity is higher in the vesicle core, water diffuses into the vesicles. The response of the vesicles to such water diffusion depends on the nature of the vesicle membranes. For vesicles with membranes in the fluid phase, formation of holes can be induced upon sudden change in osmotic pressure difference [75–78]; alternatively, the vesicles can split into smaller vesicles. Both leads to release of solutes and reduction in the osmotic pressure difference. Therefore, osmotic pressure can be used as a trigger for inducing the release of actives from these vesicles. For membranes with a gel or glassy phase, the amphiphiles that form the membranes are frozen and immobile; therefore, the membranes are not easily deformed. Thus, the membrane cannot be swollen significantly and the membrane area cannot be increased beyond the threshold imposed by the constituent amphiphiles. As a result, when subjected to a sufficiently high swelling pressure, these membranes will disintegrate and burst, providing an osmotic approach for release of actives. Similarly, when the membranes are forced to shrink in area due to diffusion of water out of the vesicles, they have to buckle and at sufficiently large reduction of the volume of the vesicle core, cracks and pores will form, leading to a release of actives, as shown in Fig. 17.

Release of actives in vesicles can also be achieved by degradation of hydrophobic blocks of amphiphiles [79]. Poly (lactic acid) and poly (caprolactone) are representative biodegradable materials due to hydrolysis of ester groups. Amphiphiles whose hydrophobic blocks are composed of polyester show increase of hydrophilic fraction due to the degradation at the end of hydrophobic block. Therefore, the amphiphiles cannot afford to support bilayer structure and generate pores on the membrane, facilitating the release of encapsulants. The amphiphiles finally disintegrate through additional degradation. The rate of degradation and time for release of actives are controllable through carefully selecting type and block length of degradable polymer or employing blends with non-degradable polymers. While non-enzymatic hydrolytic biodegradation of polyester blocks is usually slow taking place on a time scale of days to weeks [80], hydrolysis rates significantly increase in the presence of enzymes. In a similar approach, diblock copolymers with a polyester block such as poly(trimethylene carbonate)-*b*-poly (L-glutamic acid) (PTMC-*b*-PGA) degrade and release the encapsulated content within several hour when exposed to lipase enzymes [81]. The degradation of the hydrophobic PTMC block, and thus respective polymersomes, is detectable by a decrease of scattering intensity using dynamic light scattering (DLS). Manipulation of PTMC-*b*-PGA-based polymersomes is also achieved by a change of pH, which triggers a response of the polymersome within minutes due to structure transitions of PGA. As inflamed tissue as well as tissue affected by cancer exhibits acidic pH [82, 83], development of polymersomes with pH-triggered degradation

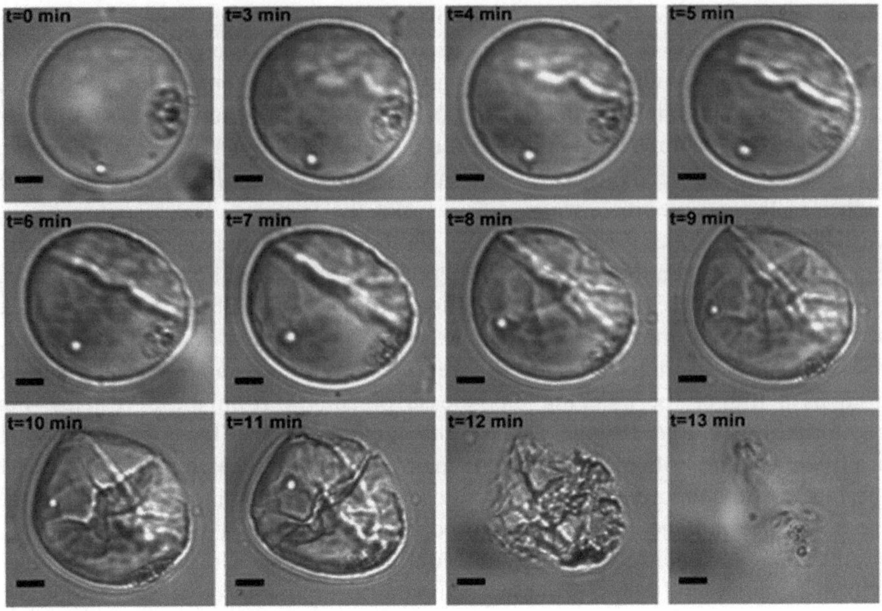

Fig. 17 Bright-field microscope images showing the shrinkage and breakage of a PEG(5000)-b-PLA(5000) polymersome after an osmotic shock. As a result of water expulsion from its inside, the polymersome shrinks and buckles. When subjected to a sufficiently high osmotic shock, the polymersome wall can break. Scale bar is 10 μm. Reproduced with permission from Ref. [63], © 2008 American Chemical Society

for localized release is the target of ongoing research. As an example, paclitaxel and doxorubicin have been successfully encapsulated and subsequently released for combined drug therapy of cancer from poly(ethylene glycol)-*block*-poly(2,4,6-trimethoxybenzylidenepentaerythritol carbonate) (PEG-*b*-PTMBPEC) by acetal hydrolysis of acid-labile PTMBPEC [84]. The degradation rates of copolymers/polymersomes are highly dependent on the molecular weight of the pH-sensitive block, but nearly unaffected by the state of aggregation, as both micellar and vesicular assemblies of PEG-b-PTMBPEC degrade within a similar time frame.

Another interesting type of polymersomes are those that combine pH-sensitivity and thermo-sensitivity. One such example has been shown with multi-responsive copolymers, double-hydrophilic poly[2-(dimethylamino)ethyl methacrylate]-*b*-poly(glutamic acid) (PDMAEMA-*b*-PGA) [85]. Thermal transition of vesicle building blocks can strongly affect self-assembly-and-degradation or perturb membrane permeability, as demonstrated in 1,2-palimitoyl-*sn*-glycero-3-phosphocholine/MPPC/DSPE/PEG liposomes [86]. In case of PDMAEMA-*b*-PGA, PDMAEMA shows LCST behavior and drives the vesicular self-assembly of the copolymer above a narrow temperature range at approximately 40 °C. Yet, most studies on thermally degradable copolymers have been utilizing *N*-isopropylacrylamide (NIPAAm) with a broad range of LCST ranging from 30 to 50 °C.

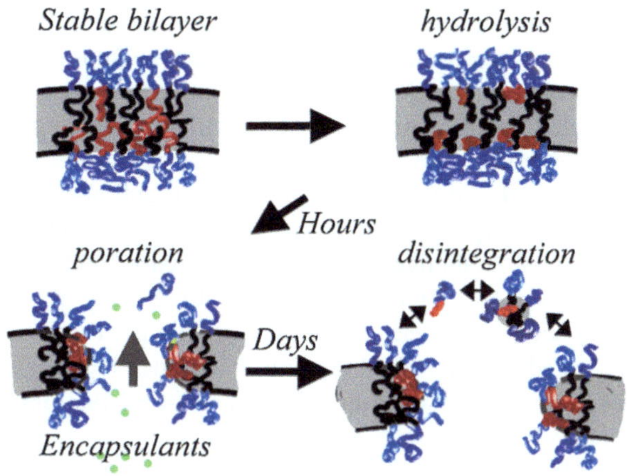

Fig. 18 Schematics of biodegradation-induced release of encapsulants. Reproduced with permission from Ref. [79], © 2004 Elsevier

Polymersomes that are made up of copolymers with a NIPAAm block, for example, poly(ethylene oxide)-*block*-poly(*N*-isopropylacrylamide) (PEO-*b*-PNIPAAm), therefore show a phase transition at a broad temperature range [87, 88]. In contrast to earlier approaches with NIPAAm-based copolymers, reversible addition–fragmentation chain transfer (RAFT) enables the fabrication of thermo-sensitive vesicular building blocks with narrow polydispersity, and thus achieves sharp phase transitions.

To degrade polymersomes in a controlled manner, one can also utilize oxidative and reductive stress, as commonly found in living tissue. these polymersomes can be fabricated from poly(propylene sulfide) (PPS) and PEG. Examples includes PEG-PSS-PEG where a disulfide (-S-S-) symmetrically divides the PSS block [89], as well as PEG-SS-PPS [90]. Similar to thiolytic cleavage of lipid membranes [91], these copolymers degrade due to formation of sulfones and sulfoxides from the thioether in the presence of oxidizing reagents such as hydrogen peroxide. For unilamellar vesicles, membrane proteins such as α-hemolysin can be incorporated for controlled release [68, 69]. This protein is attached to the bilayers and creates nanopores of approximately 2.6 nm in a diameter through the membrane due to their unique structure [92]. Therefore, encapsulated ingredients, which have smaller size than 2.6 nm, can diffuse out through these nanopores, achieving sustained release until complete depletion of the ingredients (Fig. 18).

7 Outlook

While vesicles have long been recognized as an important vehicle for delivery of drugs, cells, DNA, proteins, colorings, flavors, and other active ingredients in different industries, the widespread use of vesicles have been limited by the lack of a robust and high-throughput technique for their fabrication. Microfluidics has demonstrated great potential as a tool for fabricating liposomes, polymersomes, noisomes and other vesicular structures. Taking advantage of the high degree of control over the flow path of the fluids involved, microfluidic technologies have enabled robust fabrication of previously challenging structures, including multi-compartment vesicles. Microfluidics has opened the door towards engineering of vesicles on the individual leaflet level, and artificial cells. Nevertheless, there are still challenges towards a microfluidic-based plant for fabricating complex vesicles. One challenge is on how to increase the throughput of microfluidic approaches for fabrication. For applications that require large volume of vesicle suspensions for encapsulation, such as food and beverages, the production rates afforded by microfluidics remain insufficient. Progress has been made in the area of microfluidic scale-up [93] and this will facilitate the use of microfluidic techniques for large-scale vesicle production.

In addition, size reduction of vesicles still remains one of challenges; submicron-sized vesicles are widely useful for in vivo delivery of drugs and nutrients. Recent advances in nanofluidics enable the production of monodisperse emulsion drops whose size is less than 1 μm [94] and the principle of nanofluidics can be applied to the double emulsion drops as well. Another barrier toward practical usage of vesicles is lying in biocompatibility of materials which are used for vesicle formation in microfluidics. Although biocompatible amphiphiles have been employed, organic solvents for the amphiphiles are often harmful to human body; despite that most of the solvents are removed during vesicle formation, very small amount of the solvents may be sufficient to jeopardize in vivo usage. Therefore, additional studies on biocompatible oil phase are necessary to obviate any side effects of drug-loaded vesicles.

For further enhancing the versatility and capability of the vesicles, the characteristics and structures of the vesicles requires further manipulation. Chemical modification of amphiphiles to achieve desired functionalities will enable effective and precise delivery of vesicles to targets, while control of the number of bilayers that form the vesicle membranes will facilitate modulation of stability and permeability of vesicles.

References

1. New, R.R.C.: Liposomes : A Practical Approach. IRL Press, Oxford (1990). (Oxford University Press)
2. Discher, B.M., et al.: Polymersomes: Tough vesicles made from diblock copolymers. Science **284**, 1143–1146 (1999)
3. Zhang, L., Eisenberg, A.: Multiple morphologies of "Crew-Cut" aggregates of polystyrene-b-poly(acrylic acid) block copolymers. Science **268**, 1728–1731 (1995)
4. Munoz, S., et al.: Ultrathin monolayer lipid-membranes from a new family of crown ether-based bolar-amphiphiles. J. Am. Chem. Soc. **115**, 1705–1711 (1993)
5. Schreier, H., Bouwstra, J.: Liposomes and niosomes as topical drug carriers—dermal and transdermal drug-delivery. J. Control. Release **30**, 1–15 (1994)

6. Dinsmore, A.D., et al.: Colloidosomes: selectively permeable capsules composed of colloidal particles. Science **298**, 1006–1009 (2002)
7. Hsu, M.F., et al.: Self-assembled shells composed of colloidal particles: fabrication and characterization. Langmuir **21**, 2963–2970 (2005)
8. Segota, S., Tezak, D.: Spontaneous formation of vesicles. Adv. Colloid Interface Sci. **121**, 51–75 (2006)
9. Antonietti, M., Förster, S.: Vesicles and liposomes: a self-assembly principle beyond lipids. Adv. Mater. **15**, 1323–1333 (2003)
10. Lasic, D.D.: The mechanism of vesicle formation. Biochem. J. **256**, 1–11 (1988)
11. Kita-Tokarczyk, K., et al.: Block copolymer vesicles—using concepts from polymer chemistry to mimic biomembranes. Polymer **46**, 3540–3563 (2005)
12. Wang, Z.G.: Curvature instability of diblock copolymer bilayers. Macromolecules **25**, 3702–3705 (1992)
13. Marrink, S.J., Mark, A.E.: Molecular dynamics simulation of the formation, structure, and dynamics of small phospholipid vesicles. J. Am. Chem. Soc. **125**, 15233–15242 (2003)
14. Uneyama, T.: Density functional simulation of spontaneous formation of vesicle in block copolymer solutions. J. Chem. Phys. **126**, 114902 (2007)
15. Yamamoto, S., et al.: Dissipative particle dynamics study of spontaneous vesicle formation of amphiphilic molecules. J. Chem. Phys. **116**, 5842–5849 (2002)
16. Noguchi, H., Takasu, M.: Self-assembly of amphiphiles into vesicles: a Brownian dynamics simulation. Phys. Rev. E **64**, 041913 (2001)
17. Du, J., O'Reilly, R.K.: Advances and challenges in smart and functional polymer vesicles. Soft Matter **5**, 3544–3561 (2009)
18. He, X.H., Schmid, F.: Dynamics of spontaneous vesicle formation in dilute solutions of amphiphilic diblock copolymers. Macromolecules **39**, 2654–2662 (2006)
19. Rank, A., et al.: Preparation of monodisperse block copolymer vesicles via a thermotropic cylinder-vesicle transition. Langmuir **25**, 1337–1344 (2009)
20. Discher, D.E., Ahmed, F.: Polymersomes. Ann. Rev. Biomed. Eng. **8**, 323–341 (2006)
21. Israelachvili, J.N., et al.: Theory of self-assembly of hydrocarbon amphiphiles into micelles and bilayers. J. Chem. Soc. Faraday Trans. Ii **72**, 1525–1568 (1976)
22. Won, Y.Y., et al.: Cryogenic transmission electron microscopy (cryo-TEM) of micelles and vesicles formed in water by polyethylene oxide)-based block copolymers. J. Phys. Chem. B **106**, 3354–3364 (2002)
23. Hyde, S.T.: Curvature and the global structure of interfaces in surfactant-water systems. J. De Phys. **51**, C7209–C7228 (1990)
24. Bates, F.S., Fredrickson, G.H.: Block copolymer thermodynamics—theory and experiment. Ann. Rev. Phys. Chem. **41**, 525–557 (1990)
25. Bates, F.S.: Polymer–polymer phase-behavior. Science **251**, 898–905 (1991)
26. Bermudez, H., et al.: Molecular weight dependence of polymersome membrane structure, elasticity, and stability. Macromolecules **35**, 8203–8208 (2002)
27. Dobereiner, H.G., et al.: Mapping vesicle shapes into the phase diagram: a comparison of experiment and theory. Phys. Rev. E **55**, 4458–4474 (1997)
28. Mui, B.L.S., et al.: Influence of transbilayer area asymmetry on the morphology of large unilamellar vesicles. Biophys. J. **69**, 930–941 (1995)
29. Storm, G., Crommelin, D.J.A.: Liposomes: quo vadis? Pharm. Sci. Technol. Today **1**, 19–31 (1998)
30. Angelova, M.I., Dimitrov, D.S.: Liposome electroformation. Faraday Discuss. Chem. Soc. **81**, 303–311 (1986)
31. Sun, B.Y., Chiu, D.T.: Determination of the encapsulation efficiency of individual vesicles using single-vesicle photolysis and confocal single-molecule detection. Anal. Chem. **77**, 2770–2776 (2005)
32. Howse, J.R., et al.: Templated formation of giant polymer vesicles with controlled size distributions. Nat. Mater. **8**, 507–511 (2009)

33. Taylor, P., et al.: Fabrication of 2D arrays of giant liposomes on solid substrates by microcontact printing. Phys. Chem. Chem. Phys. **5**, 4918–4922 (2003)
34. Evans, E., Needham, D.: Physical properties of surfactant bilayer membranes: thermal transitions, elasticity, rigidity, cohesion and colloidal interactions. J. Phys. Chem. **91**, 4219–4228 (1987)
35. Mui, B., et al.: Extrusion technique to generate liposomes of defined size. Liposomes Pt A **367**, 3–14 (2003)
36. Macdonald, R.C., et al.: Small-volume extrusion apparatus for preparation of large, unilamellar vesicles. Biochim. Biophys. Acta **1061**, 297–303 (1991)
37. Frisken, B.J., et al.: Studies of vesicle extrusion. Langmuir **16**, 928–933 (2000)
38. Pautot, S., et al.: Production of unilamellar vesicles using an inverted emulsion. Langmuir **19**, 2870–2879 (2003)
39. Pautot, S., et al.: Engineering asymmetric vesicles. Proc. Natl. Acad. Sci. USA **100**, 10718–10721 (2003)
40. Shah, R.K., et al.: Designer emulsions using microfluidics. Mater. Today **11**, 18–27 (2008)
41. Squires, T.M., Quake, S.R.: Microfluidics: fluid physics at the nanoliter scale. Rev. Mod. Phys. **77**, 977–1026 (2005)
42. Brody, J.P., et al.: Biotechnology at low Reynolds numbers. Biophys. J. **71**, 3430–3441 (1996)
43. Knight, J.B., et al.: Hydrodynamic focusing on a silicon chip: mixing nanoliters in microseconds. Phys. Rev. Lett. **80**, 3863–3866 (1998)
44. Beebe, D.J., et al.: Physics and applications of microfluidics in biology. Ann. Rev. Biomed. Eng. **4**, 261–286 (2002)
45. Gambin, Y., et al.: Ultrafast microfluidic mixer with three-dimensional flow focusing for studies of biochemical kinetics. Lab Chip **10**, 598–609 (2010)
46. Pollack, L., et al.: Compactness of the denatured state of a fast-folding protein measured by submillisecond small-angle x-ray scattering. Proc. Natl. Acad. Sci. USA **96**, 10115–10117 (1999)
47. Lipman, E.A., et al.: Single-molecule measurement of protein folding kinetics. Science **301**, 1233–1235 (2003)
48. Koester, S., et al.: Visualization of flow-aligned type I collagen self-assembly in tunable pH gradients. Langmuir **23**, 357–359 (2007)
49. Koester, S., et al.: An in situ study of collagen self-assembly processes. Biomacromolecules **9**, 199–207 (2008)
50. Yun, J., et al.: Continuous production of solid lipid nanoparticles by liquid flow-focusing and gas displacing method in microchannels. Chem. Eng. Sci. **64**, 4115–4122 (2009)
51. Jahn, A., et al.: Microfluidic mixing and the formation of nanoscale lipid vesicles. ACS Nano **4**, 2077–2087 (2010)
52. Jahn, A., et al.: Controlled vesicle self-assembly in microfluidic channels with hydrodynamic focusing. J. Am. Chem. Soc. **126**, 2674–2675 (2004)
53. Jahn, A., et al.: Microfluidic directed formation of liposomes of controlled size. Langmuir **23**, 6289–6293 (2007)
54. Massignani, M., et al.: Controlling cellular uptake by surface chemistry, size, and surface topology at the nanoscale. Small **5**, 2424–2432 (2009)
55. Gullotti, E., Yeo, Y.: Extracellularly activated nanocarriers: a new paradigm of tumor targeted drug delivery. Mol. Pharm. **6**, 1041–1051 (2009)
56. Hong, J.S., et al.: Microfluidic directed self-assembly of liposome-hydrogel hybrid nanoparticles. Langmuir **26**, 11581–11588 (2010)
57. Seiffert, S., et al.: Smart microgel capsules from macromolecular precursors. J. Am. Chem. Soc. **132**, 6606–6609 (2010)
58. Thiele, J., et al.: Preparation of monodisperse block copolymer vesicles via flow focusing in microfluidics. Langmuir **26**, 6860–6863 (2010)
59. Brown, L., et al.: Polymersome production on a microfluidic platform using pH sensitive block copolymers. Lab Chip **10**, 1922–1928 (2010)

60. Karnik, R., et al.: Microfluidic platform for controlled synthesis of polymeric nanoparticles. Nano Lett. **8**, 2906–2912 (2008)
61. Kolishetti, N. et al.: Engineering of self-assembled nanoparticle platform for precisely controlled combination drug therapy. Proceedings of the National Academy of Sciences of the United States of America, vol. 107, pp. 17939–17944. 19 Oct 2010 (2010)
62. Tan, Y.C., et al.: Controlled microfluidic encapsulation of cells, proteins, and microbeads in lipid vesicles. J. Am. Chem. Soc. **128**, 5656–5658 (2006)
63. Shum, H.C., et al.: Microfluidic fabrication of monodisperse biocompatible and biodegradable polymersomes with controlled permeability. J. Am. Chem. Soc. **130**, 9543–9549 (2008)
64. Shum, H.C., et al.: Double emulsion templated monodisperse phospholipid vesicles. Langmuir **24**, 7651–7653 (2008)
65. Hayward, R.C., et al.: Dewetting Instability during the Formation of polymersomes from block-copolymer-stabilized double emulsions. Langmuir **22**, 4457–4461 (2006)
66. Shum, H.C., et al.: Dewetting-induced membrane formation by adhesion of amphiphile-laden interfaces. J. Am. Chem. Soc. **133**, 4420–4426 (2011)
67. Funakoshi, K., et al.: Formation of giant lipid vesiclelike compartments from a planar lipid membrane by a pulsed jet flow. J. Am. Chem. Soc. **129**, 12608 (2007)
68. Stachowiak, J.C., et al.: Unilamellar vesicle formation and encapsulation by microfluidic jetting. Proc. Natl. Acad. Sci. USA **105**, 4697–4702 (2008)
69. Ota, S., et al.: Microfluidic formation of monodisperse, cell-sized, and unilamellar vesicles. Angew. Chem. Int. Ed. **48**, 6533–6537 (2009)
70. Beales, P.A., et al.: Specific adhesion between DNA-functionalized "Janus'' vesicles: size-limited clusters. Soft Matter **7**, 1747–1755 (2011)
71. Shum, H.C., et al.: Multicompartment polymersomes from double emulsions. Angew. Chem. Int. Ed. **50**, 1648–1651 (2011)
72. Kisak, E.T., et al.: The vesosome—A multicompartment drug delivery vehicle. Curr. Med. Chem. **11**, 199–219 (2004)
73. Kim, S.H., et al.: Multiple polymersomes for programmed release of multiple components. J. Am. Chem. Soc. **133**, 15165–15171 (2011). doi:10.1021/ja205687k
74. Onaca, O., et al.: Stimuli-responsive polymersomes as nanocarriers for drug and gene delivery. Macromol. Biosci. **9**, 129–139 (2009)
75. Brochard-Wyart, F., et al.: Transient pores in stretched vesicles: role of leak-out. Physica A **278**, 32–51 (2000)
76. Karatekin, E., et al.: Transient pores in vesicles. Polym. Int. **52**, 486–493 (2003)
77. Karatekin, E., et al.: Cascades of transient pores in giant vesicles: line tension and transport. Biophys. J. **84**, 1734–1749 (2003)
78. Sandre, O., et al.: Dynamics of transient pores in stretched vesicles. Proc. Natl. Acad. Sci. **96**, 10591 (1999)
79. Ahmed, F., Discher, D.E.: Self-porating polymersomes of PEG-PLA and PEG-PCL: hydrolysis-triggered controlled release vesicles. J. Controlled Release **96**, 37–53 (2004)
80. Zhang, Z., et al.: The in vivo and in vitro degradation behavior of poly(trimethylene carbonate). Biomaterials **27**, 1741–1748 (2006)
81. Sanson, C., et al.: Biocompatible and Biodegradable Poly(trimethylene carbonate)-b-Poly (L-glutamic acid) Polymersomes: size control and stability. Langmuir **26**, 2751–2760 (2010)
82. Gallagher, F.A., et al.: Magnetic resonance imaging of pH in vivo using hyperpolarized C-13-labelled bicarbonate. Nature **453**, 940–973 (2008)
83. Gerweck, L.E., Seetharaman, K.: Cellular pH gradient in tumor versus normal tissue: Potential exploitation for the treatment of cancer. Cancer Res. **56**, 1194–1198 (1996)
84. Chen, W., et al.: pH-Sensitive degradable polymersomes for triggered release of anticancer drugs: A comparative study with micelles. J. Controlled Release **142**, 40–46 (2010)
85. Agut, W., et al.: pH and temperature responsive polymeric micelles and polymersomes by self-assembly of Poly 2-(dimethylamino)ethyl methacrylate -b-Poly(glutamic acid) double hydrophilic block copolymers. Langmuir **26**, 10546–10554 (2010)

86. Needham, D., Dewhirst, M.W.: The development and testing of a new temperature-sensitive drug delivery system for the treatment of solid tumors. Adv. Drug Deliv. Rev. **53**, 285–305 (2001)
87. Hong, C.Y., et al.: Synthesis and characterization of well-defined diblock and triblock copolymers of poly(N-isopropylacrylamide) and poly(ethylene oxide). J. Polym. Sci. Part A-Polym. Chem. **42**, 4873–4881 (2004)
88. Qin, S., et al.: Temperature-controlled assembly and release from polymer vesicles of poly(ethylene oxide)-block-poly(N-isopropylacrylamide). Adv. Mater. **18**, 2905 (2006)
89. Napoli, A., et al.: Glucose-oxidase based self-destructing polymeric vesicles. Langmuir **20**, 3487–3491 (2004)
90. Cerritelli, S., et al.: PEG-SS-PPS: reduction-sensitive disulfide block copolymer vesicles for intracellular drug delivery. Biomacromolecules **8**, 1966–1972 (2007)
91. Kuai, R., et al.: Efficient delivery of payload into tumor cells in a controlled manner by TAT and thiolytic cleavable PEG Co-modified liposomes. Mol. Pharm. **7**, 1816–1826 (2010)
92. Song, L., et al.: Structure of staphylococcal α-Hemolysin, a heptameric transmembrane pore. Science **274**, 1859–1865 (1996)
93. Nisisako, T., Torii, T.: Microfluidic large-scale integration on a chip for mass production of monodisperse droplets and particles. Lab Chip **8**, 287–293 (2008)
94. Malloggi, F., et al.: Monodisperse colloids synthesized with nanofluidic technology. Langmuir **26**, 2369–2373 (2010)

Dielectrophoresis Field-Flow Fractionation for Continuous-Flow Separation of Particles and Cells in Microfluidic Devices

Nuttawut Lewpiriyawong and Chun Yang

Abstract The ability of separating particles and cells in continuous flow is always desirable as it allows faster biomedical diagnosis. This chapter presents a review on the development of dielectrophoresis field-flow fractionation (DEP-FFF) technologies for continuous-flow separation of particles and cells in microfluidic devices. The review is mainly focused on the publications between 2005 and 2012. During separation processes, DEP-FFF transports particles and cells with hydrodynamic liquid flow in microchannels and fractionates particles and cells using dielectrophoresis (DEP) force generated perpendicular to the fluid flow direction. In the literature, numerous strategies have been developed to advance the way of generating nonuniform electric field which is required to produce DEP force, and four main strategies are grouped here, including (1) the use of planar electrodes, (2) the use of three-dimensional and sidewall electrodes, (3) the use of insulating topographical structures, and (4) the use of combined planar electrodes and insulating structures. DEP-FFF microfluidic devices can serve as a label-free, non-invasive and most efficient tool for manipulating and separating various biosamples such as particles, cells and DNA based on their polarizabilities in nonuniform electric field.

1 Introduction

Separation of target cells is of broad importance in a large number of applications. In biomedical diagnoses, separation of platelets, white or red blood cells from blood is a common practice in modern medical laboratories and hospitals [1]. To enhance the effectiveness of artificial fertilization, healthy sperm cells must be separated from the unhealthy ones [2]. In environmental monitoring, diagnosis of

N. Lewpiriyawong · C. Yang (✉)
School of Mechanical and Aerospace Engineering, Nanyang Technological University, Nanyang Avenue 639798, Singapore
e-mail: mcyang@ntu.edu.sg

L. Wang (ed.), *Advances in Transport Phenomena 2011*,
Advances in Transport Phenomena 3, DOI: 10.1007/978-3-319-01793-8_2,

water quality involves identification, separation and counting of live bacterial cells [3]. For these clinical and industrial applications, centrifugation, chromatography, flow cytometry and fluorescence/magnetic-activated cell sorters (FACS/MACS) are the leading separation and detection techniques [4]. However, these existing macro-scale techniques have the following drawbacks: (1) non-continuous separation process which leads to longer analysis time, (2) requirement of a large volume and modification of samples, and (3) the need for highly trained personnel to operate bulky and costly instruments [4–8].

Lab-on-a-chip (LOC) or micro-total analysis system (μTAS) is a fast-emerging interdisciplinary technology that integrates laboratory based biomedical and chemical processes onto a single chip with typical size of only millimeters to centimeters [9]. In LOC platforms, microfluidics, consisting of capillary network structures, exploits the characteristics of laminar flow to provide precise control of fluid in nanolitres and to manipulate thousands of cells through either pressure gradient or electrokinetic forces. Thus, microfluidic devices have numerous advantages such as fast analysis, requirement of small amount of sample volume, portability, disposability, etc. These benefits are favorable to the development of future inexpensive clinical and point-of-care diagnostic devices. It is therefore unsurprising that microfluidics has been utilized in manipulation, separation and identification of cells, proteins and DNA, drug delivery, and rapid diagnosis of complex diseases [10].

One of the most important clinical and industrial demands in biomedical diagnosis is the separation of target biosamples in a continuous manner. Field-Flow Fractionation (FFF) was engineered using the principle similar to chromatography with the aim to achieve continuous-flow separation of subpopulations without stationary phase [11]. FFF transports the samples by the means of pressure-driven flow in microchannels and fractionates them using an externally applied field perpendicular to the flow direction. FFF subtechniques are mainly classified by the externally applied field: sedimentation (SdFFF), thermal (ThFFF), flow (FlFFF), magnetic (MgFFF), electrical (ElFFF) and dielectrophoresis (DEP-FFF) [12]. Among these subtechniques, DEP-FFF is the most promising technique that has exploited both microfabrication and microfluidic technologies for continuous-flow separation of various samples such as human leukemia (HL-60) cells [13], breast cancer cells [14, 15], bacteria [16], yeast cells [17] and DNA [18].

DEP is an electrokinetic phenomenon describing the motion of a dielectric particle or cell in nonuniform electric field due to polarization effects [19]. A DEP force is resulted from the nonuniform polarization effects generated by the interaction of induced dipoles with the externally applied AC or DC electric field gradients, thus allowing separation of particles and cells based on their different polarizabilities to their suspending medium [20]. Using this concept, the DEP technique unlike FACS and MACS does not need any pre-treatment to samples such that the sample properties are maintained for further downstream processes. This attractive advantage makes DEP a non-invasive and simple yet effective tool which is uniquely featured for continuous-flow separation of target cells from other interfering particulates in microfluidic devices.

This review article presents advancement of the development of state-of-the-art DEP-FFF technologies for continuous-flow separation of biosamples in microfluidic devices. The publications between 2005 and 2012 are the main focus of this review. The article classifies DEP devices into four groups based on the strategies of generating nonuniform electric field for inducing DEP force. These four strategies are: (1) the use of planar (or thin film) electrodes, (2) the use of three-dimensional (3D) and sidewall electrodes, (3) the use of insulating topographical structures, and (4) the use of combined planar electrodes and insulating structures (known as hybrid DEP system). In the review, the theoretical background of DEP is provided and followed by the separation mechanisms and fabrication techniques of the four DEP groups. Finally, the summary section provides possible future research directions for improving DEP technologies in the field of continuous separation of biosamples.

2 Theoretical Background of Dielectrophoresis

DEP refers to the motion of a polarizable particle (or a cell) in a buffer solution under nonuniform electric field due to polarization effects. The effects occur as a result of the interaction between the unevenly distributed induced dipoles in the particle and the applied nonuniform electric field [20]. As shown in Fig. 1a and b, the interaction between the induced net dipole charges and the nonuniform electric field leads to a nonzero columbic net force (F_{DEP}) on the particle [21, 22]. The DEP force can be determined by using the Maxwell stress tensor formulation [23] or the effective dipole moment method [21]. Both methods have been shown to yield an identical DEP force expression. Furthermore, because the effective dipole moment method is much simpler, it has been widely used in the DEP community [24]. The method points out that the effective dipole moment of a neutral particle

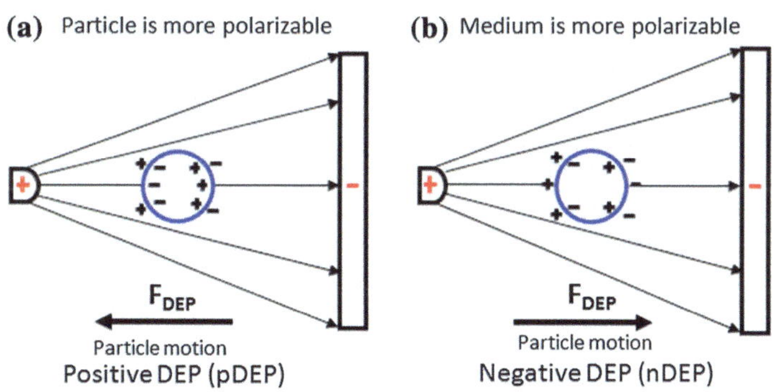

Fig. 1 Illustration of dielectrophoretic (*DEP*) force. **a** Positive DEP (*pDEP*) attracting the particle to the high electric field region and **b** Negative DEP (*nDEP*) repelling the particle away from the high electric field region [22]

polarized by an electric field is defined as the moment of an equivalent free-charge point dipole that is immersed in the same dielectric medium and located at the same point of the particle's center [21]. With the dipole moment method, the time-average DEP force acting on a spherical homogeneous particle of radius a suspended in a dielectric medium with permittivity ε_m is expressed as [21]

$$\langle \vec{F}_{DEP} \rangle = 2\pi\varepsilon_m a^3 \mathrm{Re}\left[\underline{f}_{CM}(\omega)\right] \nabla \vec{E}^2_{rms,ac} \tag{1}$$

where ∇ denotes a gradient operation, $\vec{E}^2_{rms,ac}$ is the square of the root-mean-square of local AC electric field and $\mathrm{Re}\left[\underline{f}_{CM}(\omega)\right]$ is the real part of the Clausius–Mossotti (CM) factor given by

$$\underline{f}_{CM}(\omega) = \frac{\underline{\varepsilon}_p - \underline{\varepsilon}_m}{\underline{\varepsilon}_p + 2\underline{\varepsilon}_m} \tag{2}$$

where $\underline{\varepsilon}_p$ and $\underline{\varepsilon}_m$ are the complex permittivities of the particle and suspending medium, respectively. The complex permittivity is represented by $\underline{\varepsilon} = \varepsilon - i\sigma/\omega$ where ε and σ are respectively the permittivity and electrical conductivity, $i = \sqrt{-1}$, and $\omega = 2\pi f$ is the radian field frequency. Therefore, $\underline{f}_{CM}(\omega)$ depends on electric field frequency and complex permittivities of the particle and suspending medium. Moreover, $\underline{f}_{CM}(\omega)$ provides a measure of effective polarization strength, and the value of its real part varies from -0.5 to 1. When the particle is more polarizable than the suspending medium (i.e., $\mathrm{Re}\left[\underline{f}_{CM}(\omega)\right] > 0$), positive DEP (pDEP) is generated to attract the particle to the high electric field region as illustrated in Fig. 1a. When the particle is less polarizable than the suspending medium (i.e., $\mathrm{Re}\left[\underline{f}_{CM}(\omega)\right] < 0$), negative DEP (nDEP) is generated to repel the particle away from the high electric field region as illustrated in Fig. 1b. At high frequency (e.g., $\omega \to \infty$), the CM factor reduces to $\mathrm{Re}\left[\underline{f}_{CM}(\omega)\right] = (\varepsilon_p - \varepsilon_m)/(\varepsilon_p + 2\varepsilon_m)$ which is only dependent on the permittivities of the particle and suspending medium. On the other hand, at low frequency (e.g., $\omega \to 0$), $\mathrm{Re}\left[\underline{f}_{CM}(\omega)\right] = (\sigma_p - \sigma_m)/(\sigma_p + 2\sigma_m)$ which is only dependent on the electrical conductivities of the particle and suspending medium where the particle conductivity σ_p is the sum of its bulk conductivity $(\sigma_{p,bulk})$ and its surface conductivity $(2Ks/a)$. For insulating latex particles ranging from micron to submicron in diameter, $\sigma_{p,bulk}$ is typically negligible compared to the conductivity of suspending medium σ_m, and Ks is the surface conductance ranging from 0.2 to 2.1 nS [25, 26]. In particular, for an applied DC electric field, the time-average DC-DEP force expression is the same as that presented in Eq. 1 as $\vec{E}_{rms} = \vec{E}_{dc}$.

An AC electric field can be biased with a DC electric field. Under an assumption that the AC electric field frequency is low (e.g., $\leq$10 kHz), then $\underline{f}_{CM,ac} \approx \underline{f}_{CM,dc}$, and thus the time-average DEP force can be formulated as [27]

$$\langle \vec{F}_{DEP} \rangle = 2\pi\varepsilon_m a^3 \mathrm{Re}\left[\underline{f}_{CM}(\omega)\right] \nabla \left(\vec{E}^2_{dc}(\vec{r}) + \vec{E}^2_{rms,ac}(\vec{r}) \right) \tag{3}$$

If there are spatial changes in the phase $(\varphi_x, \varphi_y, \varphi_z)$ of an AC electric field with their respective magnitudes (E_{x0}, E_{y0}, E_{z0}), the generalized DEP force can be expressed as [28],

$$\langle \vec{F}_{DEP} \rangle = 2\pi\varepsilon_m a^3 \left\{ \begin{array}{l} \mathrm{Re}\left[\underline{f}_{CM}(\omega)\right] \nabla \vec{E}^2_{rms,ac} \\ +\mathrm{Im}\left[\underline{f}_{CM}(\omega)\right] \left(E^2_{xo} \nabla \varphi_x + E^2_{yo} \nabla \varphi_y + E^2_{zo} \nabla \varphi_z \right) \end{array} \right\} \tag{4}$$

where $\mathrm{Im}\left[\underline{f}_{CM}(\omega)\right]$ is the imaginary part (or the out-of-phase component) of the CM factor and it indicates whether the induced dipole moment leads $\left(\mathrm{Im}\left[\underline{f}_{CM}(\omega)\right] > 0\right)$ or lags $\left(\mathrm{Im}\left[\underline{f}_{CM}(\omega)\right] < 0\right)$ the applied electric field. When the induced dipole moment leads the electric field, the particle moves in the direction opposing to that of the traveling field or vice versa [28].

3 DEP Devices

3.1 Planar (or Thin-Film) Electrodes

With the advancement of microfabrication technologies, planar (or thin-film) electrodes fabricated by using the sputtering and lift-off processes have been extensively integrated in DEP devices [29]. Due to the miniaturized dimensions of microfabricated electrodes, an electric field gradient and the resulting DEP force are significantly enhanced for facilitating manipulations of micro-sized particles and biological cells.

3.1.1 Dielectrophoresis Field-Flow Fractionation (DEP-FFF)

DEP-FFF being one of the FFF subtechniques utilizes DEP force to separate particles and cells in a continuous flow. DEP force is generated due to the presence of nonuniform electric field which is conventionally induced by planar interdigitated electrodes deposited on the chamber floor of a microfluidic device as schematically shown in Fig. 2a. When particles are less polarizable than their suspending medium, they experience repulsive nDEP force that pushes the particles to levitate above the DEP electrodes. On the other hand, particles experiencing pDEP force can be attracted to (or trapped at) the DEP electrodes. Two DEP-FFF operation modes can be classified based on how the particles interact with DEP force. The first operation mode can be used when two different particle groups experience nDEP of difference in magnitude. Figure 2a shows that particles in the DEP-FFF channel can attain different equilibrium heights based on the

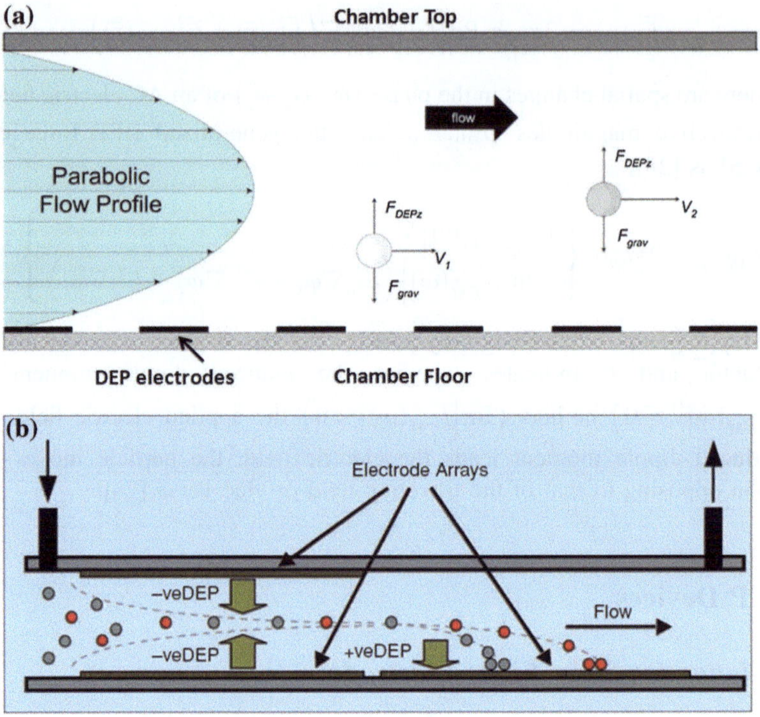

Fig. 2 Separation principle of DEP-FFF. **a** A conventional DEP-FFF device with interdigitated electrodes levitating particles by nDEP force [31]. **b** Combined pre-focusing and fractionation functions using both nDEP and pDEP [35]

balance between nDEP and gravitational forces. The particles at the higher equilibrium height having a higher velocity in the pressure-driven flow will be separated prior to the others [30, 31]. Such DEP-FFF principle has been successfully used for the separation of polystyrene latex beads [32] and human leukemia (HL-60) cells from peripheral blood mononuclear cells [13] as well as for the detection of malaria cells.

The second mode can separate particles undergoing different polarities of DEP forces, namely pDEP and nDEP. The particles experiencing nDEP are levitated above the electrodes while the others undergoing pDEP are attracted to the electrodes. Such the device utilizing the nDEP and pDEP operation mode has been widely exploited for separating various biological samples such as viable from non-viable yeast cells [17, 33], erythrocytes from latex beads [34], different types of white blood cells [35] and cultured human breast cancer M4DA-435 cells from normal blood cells [14, 15]. Continuous DEP cell separation was reported by using a series of short-time flushing; however, the system still needed certain relaxation time for the cells to reach equilibrium after each on–off flushing [36]. In addition to these typical interdigitated electrodes, other electrode designs including multilayered grid electrode array [37], ring electrodes [38] and combined quadrupole and microwell electrodes [39] were adopted to achieve trapping of microparticles and single cells in DEP-FFF microfluidic devices.

Furthermore, the DEP-FFF technique was advanced by combining pre-focusing and fractionation functions to specifically trap cells into distinct locations [35]. In Fig. 2b, the first section consisting of two interdigitated electrode arrays at the top and bottom of the channel can focus particles into the centre zone of the channel by nDEP. The second section with an electrode array differentially attracts the focused particles by pDEP into distinct bands. Other than particle separation, DEP-FFF devices have also been used for cell lysis and trapping [40], immobilization of probe beads for affinity assay [41], cell genetic detection [42] and analyses of cellular properties such as dielectric, density and deformability [43].

Despite these successes, the separation efficiency of DEP-FFF devices may be compromised due to the following issues. The first issue is related to the overlapping of sample zones which is mainly caused by similar displacement velocity and dispersion of individual subpopulations during separation [11]. However, this issue can be resolved by optimizing the chamber dimensions, applied electric field strength and flow rate. The second issue concerns the exponential decay of electric field strength and thus DEP force at a further distance from the surface of planar electrodes [32, 44]. As a result, the manipulation of cells far away from the electrode surfaces becomes ineffective.

3.1.2 Lateral DEP-FFF

Because of the rapid decay of electric field strength from the surface of conventional DEP-FFF devices, researchers have developed DEP-FFF devices to extend the DEP force across the entire channel height. Dürr et al. [45] devised a microfluidic device with interdigitated electrode pairs which are embedded on the bottom and top of the channel while being inclined with the flow direction. Such electrode configuration is also known as DEP deflector. As shown in Fig. 3a, these double-sided electrodes allow electric field to distribute across the channel height, thus mitigating the issue of exponential decay of electric field and resulting in better manipulation of cells. In Fig. 3b, as the electrodes are inclined at an angle relative to the flow direction, the DEP force acting on particles of different sizes induces distinct lateral displacement of each subpopulation, leading to separation into different outlets simultaneously [46]. The first deflector only deflects large particles to the collecting outlet whereas smaller particles are discarded at the waste outlet. The second deflector further screens out the particles larger than the target size. As a result, separation of mitochondria (empty circles) from a human lymphoblastoid cell line (solid circles) was successfully demonstrated. Furthermore, Chen et al. [47, 48] utilized these electrode deflectors as DEP barriers which allow smaller particles to penetrate, but slow down (or trap) larger particles, thus achieving particle separation and accumulation by size simultaneously.

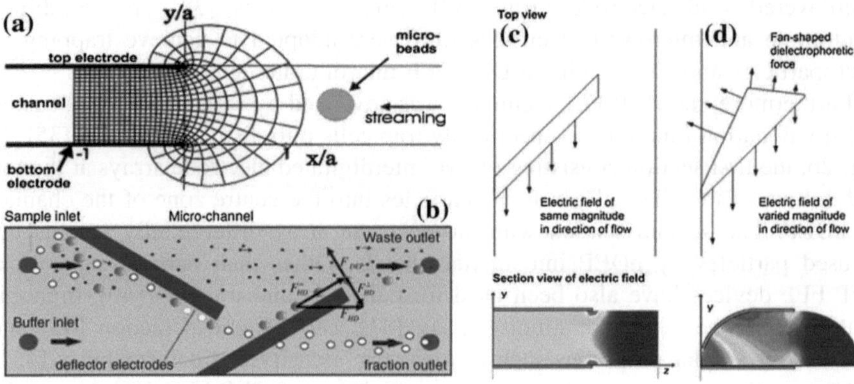

Fig. 3 DEP separation with electrode deflectors. **a** Side view of double-sided electrode deflectors generating electric field across the channel height [45]. **b** Lateral size-dependent separation using two consecutive deflectors [46]. **c** A typical deflector generating electric field of the same magnitude along the flow direction and **d** a 3D fan-shaped electrode DEP system generating electric field of varying magnitude along the flow direction in a half-circular channel [54]

Besides straight DEP electrodes, other electrode designs including trapezoidal, triangular and curved shapes were also widely employed for separation of microparticles and cancer cells [49–51] and focusing of nanoparticles [52]. Ling et al. [53] recently presented a new device with an isosceles triangular ITO electrode array to continuously separate 10 and 15 μm particles with a speed of 1 mm/s. Instead of the lift-off technique, surface micromachining with integrated fluid technology was used to fabricate DEP gating for focusing and separating *Penicillium brevicompactum* (PBC), human T-cells and *E. coli* [16]. However, disregarding the shape of these typical electrode deflectors, electric fields of the same magnitude can only be created along the flow direction as illustrated in Fig. 3c. Park et al. [54] developed a 3D asymmetric fan-shaped electrode system that can produce electric fields of continuously varying magnitudes along the flow direction due to the varying width and depth between deflectors in a half-circular channel (Fig. 3d). Hence, variation in DEP force was produced to increase the sensitivity of lateral sorting of cells. This microfluidic device showed the purities of 81.5 ± 7.6 % for mouse P19 embryonic carcinoma and of 94.1 ± 4.3 % for red blood cells while the typical deflector system could only provide the purities of 79.4 ± 6.8 % for mouse P19 embryonic carcinoma and of 91.3 ± 3.7 % for red blood cells.

It is noted that although the double-sided electrodes permit electric field to distribute along the channel height, the electric field strength is not always uniform along the height and it could vary substantially when there is a need to increase channel height for higher throughput. Therefore, DEP force magnitudes along the channel height are no longer uniform as desired. Interestingly, Liao et al. [55] recently utilized this concept to precisely separate 2, 3, 4 and 6 μm polystyrene particles by size. As illustrated in Fig. 4, because of the presence of the DEP force

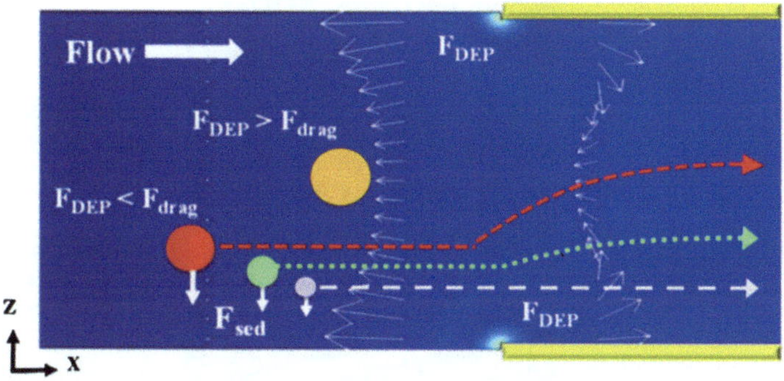

Fig. 4 Separation of polystyrene particles by size (e.g., 2, 3, 4 and 6 μm) using the DEP force gradient along the z direction generated by an inclined DEP gate [55]. When DEP force is greater than drag force, the largest particle is trapped. On the other hand, smaller particles experiencing weaker DEP forces are dominated by the drag force. Thus, these smaller particles have different penetration times and distinct lateral displacement along the inclined DEP gates

gradient from the channel top to the bottom (in the z direction), particles of different sizes have different penetration time and distinct lateral displacement along the inclined DEP gates. Therefore, it is clear that the channel height becomes the most critical parameter to generate the desirable DEP force gradient and thus to determine the system efficiency. However, upon increasing the channel height for obtaining better throughput, the system efficiency and sensitivity may be compromised since the DEP force gradient is strong only near the electrodes (on the top and the bottom) but weak at the centre. Consequently, the particles of similar size can indiscriminately penetrate these DEP gates. In fact, this mechanism proposed by Liao et al. [55] is similar to the aforementioned technique with DEP barriers proposed by Chen et al. [47] which however, did not consider the variation of DEP forces along the z direction. As the channel height is not a suitable parameter to increase the throughput, one may increase the channel width with longer electrodes. However, it is not strongly recommended for biological samples to be deflected near electrode edges in the elevated temperature zone for a long time because cells can be physiologically damaged. Additionally, cells can also collide, aggregate and clog the system especially when the sample concentration is high [46]. Alternatively, separation of samples can be performed in parallel. It was shown that cascade and staggered cell sorters allowed for an increase in the flow rates from 0.2 to 0.65 μL/min without compromising enrichment factor and meanwhile could achieve the improved isolation purity from 80 to 96 % [56].

In addition to most proof-of-concept DEP separations of a binary mixture, separation of multiple particles and biosamples are of great interest nowadays and can be achieved using planar electrodes with multiple angles of inclination. As shown in Fig. 5a, separation of 3, 5 and 10 μm microparticles was successfully demonstrated using piecewise curved electrodes, inclined at 9.5°, 14° and 18.4° in relative to the flow direction [57]. Furthermore, Kim et al. [58] reported the lateral separation of multiple bacterial cell types with the aid of distinct synthetic DEP

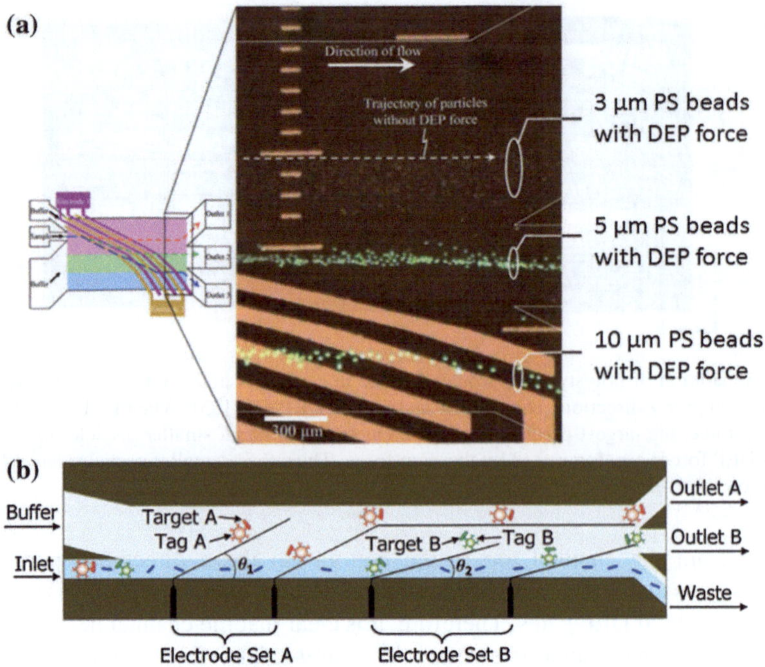

Fig. 5 DEP separation of multiple samples with inclined planar electrodes. **a** Separation of 3, 5 and 10 µm microparticles using piecewise curved electrodes inclined at 9.5°, 14° and 18.4° relative to the flow direction [57]. **b** Separation of multiple bacterial cells with the aid of distinct synthetic DEP tags in multi-target DEP activated cell sorter (*MT-DACS*) [58]

tags in multi-target DEP activated cell sorter (MT-DACS) as shown in Fig. 5b. The DEP tag A and tag B are 10 µm non-fluorescent and 5 µm fluorescent polystyrene particles coated with anti-bacteria antibodies. Because the tag A is larger than the tag B, the electrode set A inclined with a grater angle ($\theta_1 = 10°$) exerts a larger DEP force on the tag A, thus separating the target A with tag A to the outlet A. Simultaneously, the electrodes set B inclined with a smaller angle ($\theta_2 = 8°$) separate the target B with tag B to the outlet B, while allowing the untagged bacteria to separate at the outlet C. Not only for bacteria, DACS systems were also used to separate microparticles and platelets from dilute blood samples [57, 59–62].

3.1.3 DEP with Modulation in Space or in Time

In previous cases, DEP force is generated with neither modulation in space nor in time because fixed AC electric field is continuously applied to all planar electrodes in the device. This section reviews three DEP techniques which modulate DEP force in space or in time, namely, traveling wave DEP, moving DEP and pulsed

DEP. Using such modulated-DEP techniques, continuous separation of samples can be achieved without the need of fluid flow like conventional DEP-FFF.

Traveling-wave DEP (twDEP)

The working principle of twDEP is based on the application of AC electric field with a typical 90° phase shift to adjacent planar electrodes so as to modulate twDEP force in space as shown in Fig. 6a [63, 64]. According to Eq. (4), the twDEP force is dependent on the DEP force associated with both $\mathrm{Re}\left[\underline{f}_{CM}(\omega)\right]$ for nDEP levitation (or pDEP attraction) and $\mathrm{Im}\left[\underline{f}_{CM}(\omega)\right]$ for translational motion. Depending on the applied field frequency, particles could be transported in the same or opposite direction to that of the traveling wave propagation direction. The translational twDEP velocity of a particle is proportional to the square of the electric field strength and it determines the speed in separation process [65].

Similar to DEP-FFF, there are two twDEP operation modes. The first mode is used when two samples experience different types of DEP force (i.e., pDEP and nDEP) while the second mode is operated when two samples undergo the same type of DEP force (e.g., nDEP). The first mode was demonstrated to fractionate non-viable from viable yeast cells at 35 kHz [63, 65]. The non-viable cells were attracted to the electrodes by pDEP while the viable cells were levitated and eluted by twDEP force. Another important twDEP development using the first mode was the detection of malaria from human blood samples by using the spiral electrode configuration as shown in Fig. 6b [66]. Malaria cells (displayed in green at the

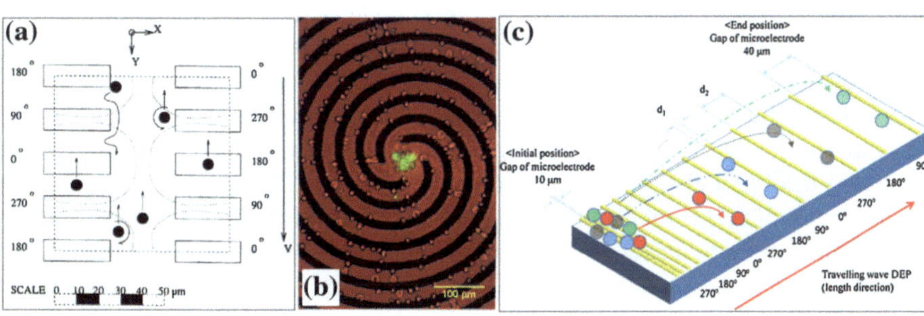

Fig. 6 twDEP electrode configurations. **a** Conventional twDEP electrodes addressed with voltages of 90° phase sequences establishing traveling wave of propagation direction. Depending on the applied field frequency, particles are transported in the same or opposite direction to that of the traveling wave [64]. **b** Spiral twDEP electrodes employed for detecting malaria-infected cells from human blood samples. The infected cells (displayed in *green* at the *centre*) are levitated by nDEP and transported towards the electrode centre by twDEP while the normal cells are trapped by pDEP at electrode edges [66]. **c** A very large area (31 × 25 mm^2) twDEP electrode array with gradually increasing gaps from 10 to 40 μm produced both nDEP and gradient twDEP forces to balance gravitational force for multiple particle separation by size [67]

centre) exhibited nDEP were levitated and transported towards the centre of the spiral twDEP electrodes while normal cells were trapped by pDEP at electrode edges. This separation platform allows a simple detection of malaria-infected cells in stagnant flow. In the second mode, as both samples undergo the same nDEP force, a difference in the twDEP translational velocity of each sample determines successful separation. Based on this concept, erythrocytes and leukocytes (both undergoing nDEP) were successfully fractionated as erythrocytes move faster than leukocytes in the travelling wave [68]. twDEP-based separation of other samples including latex beads [69] and DNA molecules by size [18] was also demonstrated.

In addition to the above typical twDEP configurations, other twDEP designs have also been reported. As illustrated in Fig. 6c, twDEP electrodes could be positioned with gradually increasing gaps from 10 to 40 μm in a very large area ($31 \times 25\ mm^2$) [67]. When the gaps are larger, nDEP and gradient twDEP forces, used for balancing gravitational force, become weaker towards the channel end, causing particles to settle at different locations. The device was successfully used to separate 3, 6, 10 and 20 μm particles. Another advanced twDEP technique is the use of twDEP force to balance gravitational force which is produced by inclining the twDEP device at an angle for separation of particles by size [70].

Other than separation in stagnant fluid, twDEP has been implemented for lateral separation of samples in continuous flows. The separation principle is based on two working steps: first, samples are focused to the centre of the separation area, and second, the focused samples are separated laterally by twDEP force. Figure 7a shows a PDMS-based twDEP device with ITO electrodes used for cellular characterization and separation of viable human myelogenous HL-60 cells from the non-viable ones [71]. The device employed hydrodynamic focusing via multiple inlets to the centre. When twDEP with 1 kHz frequency was activated, it was found that the viable cells moved perpendicularly to the flow with a faster speed (4.3 μm/s) than the non-viable cells (3 μm/s) under a flow rate of 5 ml/min. As a result, the viable cells were collected at the outlet farther away from the centre as compared to the non-viable cells. It was also reported that the use of ITO electrodes is beneficial for efficient tracking and image analysis of cell motion since they are optically transparent and provide much lower background noise as compared to Cr/Au electrodes. Subsequently, a new twDEP device was proposed to implement 3D DEP focusing and 3D twDEP force which can be induced by top and bottom electrode arrays as illustrated in Fig. 7b [72]. The device function was tested to separate red blood cells from *Staphylococcus aureus* bacterial cells with a maximum flow rate of up to 10 μL/min (or linear velocity of $\sim$3 mm/s).

It is worthwhile to note two main limitations of twDEP systems. First, higher electric field strength may speed up the twDEP separation process but it also can impose current-induced Joule heat into the medium and thus can damage the biosamples. Second, with planar electrode structure, twDEP devices still suffer the exponential decay of electric field away from the electrode surfaces, which is similar to the problem encountered in the conventional and lateral DEP-FFF systems.

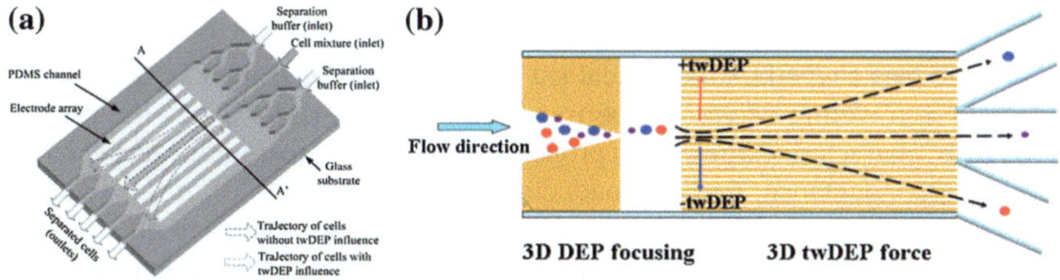

Fig. 7 Focusing-aided twDEP techniques for lateral separation. **a** Hydrodynamic focusing [71] and **b** DEP focusing [72]. twDEP electrodes produce twDEP force perpendicularly to the flow direction causing lateral separation of samples

The twDEP system not only can be used for sample separation but also can be served as a tool for characterizing dielectric properties and frequency-dependent DEP behaviors of particles and cells. After obtaining such information, researches can predict whether cells in a buffer solution experience pDEP or nDEP at a specific frequency, and can be used for performing separation. This twDEP-based characterization tool is named as electrorotation (ROT). Different from the typical twDEP systems, the ROT electrode configuration consists of two pairs of electrodes positioned perpendicularly to each other (Fig. 8). By assigning sine and cosine wave signals to individual pairs, a rotating electric field and thus a rotational torque can be generated on a polarized particle/cell [21]. Since a cosine wave always leads a sine wave by 90° of phase, the resulting electric field rotates anti-clockwise. This rotating wave can be used to levitate the particle/cell above the electrodes by repulsive DEP force and to rotate the particle/cell by the torque in the same or opposite direction to that of the rotating field. The torque exerting on a particle/cell can be expressed as [73, 74]

$$\Gamma(\omega) = -\mathrm{Im}\left[\underline{\vec{p}}_{eff}(\omega)\right]\vec{E}_{ac} \tag{5}$$

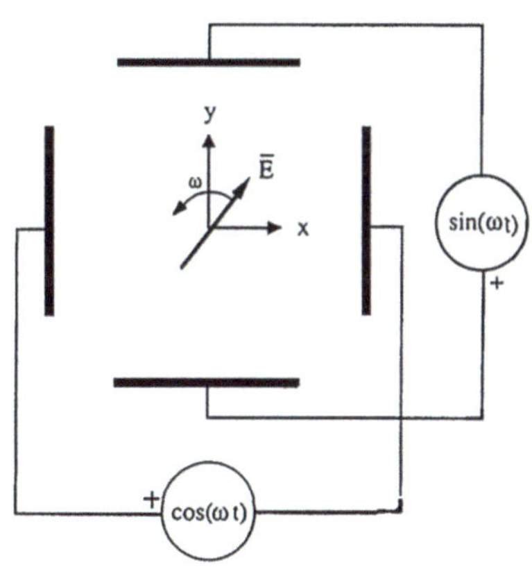

Fig. 8 Electrorotation electrode (*ROT*) configuration consisting of two pairs of perpendicular electrodes assigned with sine and cosine wave signals to induce a rotating electric field and rotational torque on a polarized particle/cell [21]

where $\vec{p}_{eff}(\omega) = 4\pi\varepsilon_m a^3 f_{CM}(\omega)\vec{E}_{ac}(\vec{r}, t)$ is the complex effective dipole moment of a dielectric particle/cell. A theoretical study of detailed electrokinetic behaviors of a particle in traveling waves can be found elsewhere [64]. ROT has been widely utilized to characterize dielectric properties (including membrane capacitance, membrane conductance, surface conductance and zeta potential) of latex spheres [26], human leukocyte subpopulations [75], cell membrane [76], and breast cancer cell lines [77]. Details of the apparatus and procedures can be found elsewhere [78]. The current theory usually extracts cell dielectric properties from the measurement of crossover frequencies, indicating the transition frequencies between pDEP and nDEP, or the measurement of peak frequencies, providing the maximum electrorotation rate [79, 80]. However, since the expression of both crossover and peak frequency are nonlinear and cannot be solved analytically, the usual practice is to employ the low frequency (DC) approximations of Schwan [81] which require the measurements to be carried out in a medium with conductivity less than 10 μS/cm. Recently, Lei et al. [80] greatly modified the ROT theory by applying various levels of approximations through scaling analysis, and they gave the full expressions of equivalent cell permittivity and conductivity. Their study showed successful extraction of both membrane and interior properties of cells in suspending media with any conductivity from the peak or crossover frequency measurements. In addition to the DEP characterization of cells, Fuhr et al. [82] demonstrated the separation of plant protoplasts using a ROT system in which cells having distinct membrane capacitances were observed to rotate in different directions and then were separated.

Moving DEP

Moving DEP (mDEP) makes use of the real part of the induced dipole moment in order to simultaneously transport and fractionate particle and cell samples along a microfluidic channel without the need of fluid flow [83]. In operation, either a single electrode or an array of electrodes is sequentially energized at a time, enabling mDEP force to be created or modulated in time. As shown in Fig. 9, the dotted circle represents a cell at the initial time ($t = 0$) prior to energizing the electrode while the patched and solid circles denote the cells after experiencing pDEP and nDEP, respectively. The pDEP cell is attracted to the high electric field region near the energized electrode and the nDEP cell is repelled away from the energized electrode. By energizing subsequent electrodes from $t = T$ to $2T$, the nDEP cell is continuously repelled to the right while retaining the pDEP cell at the first electrode. At this stage, the fractionation of two subpopulations is noticed with a distance of two electrodes apart. At $t = 3T$, the sample transport begins by moving three energized electrodes to the right, causing separated cells to transport with the moving electric field. mDEP was successfully demonstrated to fractionate viable and non-viable yeast cells. Optimization of the device configuration and studies of cell motion under mDEP, drag, buoyancy and gravitational forces were also conducted by Kua et al. [84, 85]. Clearly, the technique reduces the

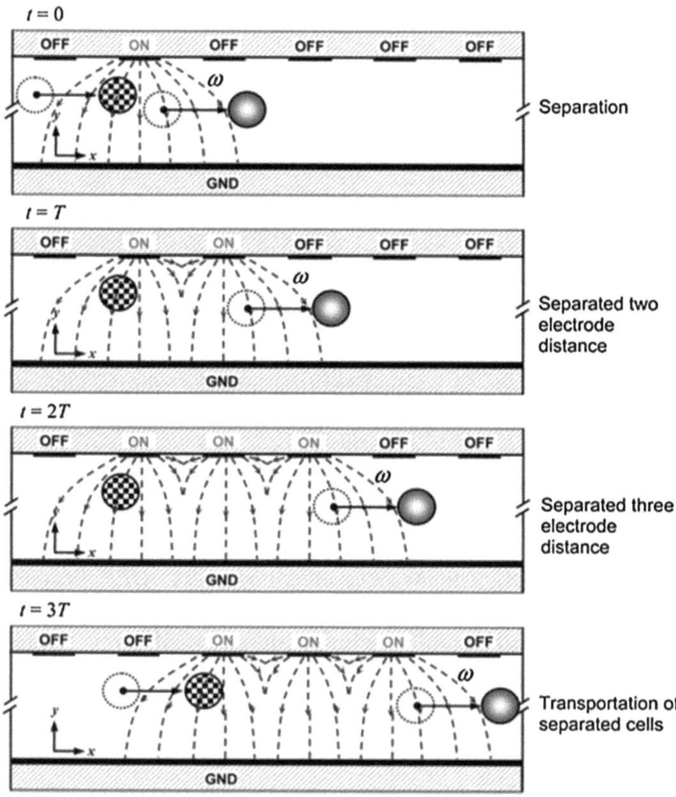

Fig. 9 Moving DEP for fractionation of cells under pDEP (*patched circle*) and nDEP (*solid circle*) [83]

complexity of the twDEP system; because the latter requires both the real and imaginary parts of the induced dipole moment. However, mDEP can only fractionate cells in a stationary fluid, and thus suffers low throughput. Furthermore, similar to twDEP, additional improvement in terms of flow-through separation and the decay of electric fields from planar electrode surface is still needed.

Pulsed DEP

Pulsed DEP is the most recent developed technique which employs DEP force with modulation in time for particle separation by size in a DEP-FFF microchannel as depicted in Fig. 10a [86]. In this technique, nDEP force is pulsed in time (or discontinuous) while the frequency of pulsing can be controlled. The suitable pulsing frequency allows the system to reverse the order of separation (That is, eluting the larger particles prior to the smaller ones, and also extracting intermediate-size particles from a heterogeneous sample in a single run).

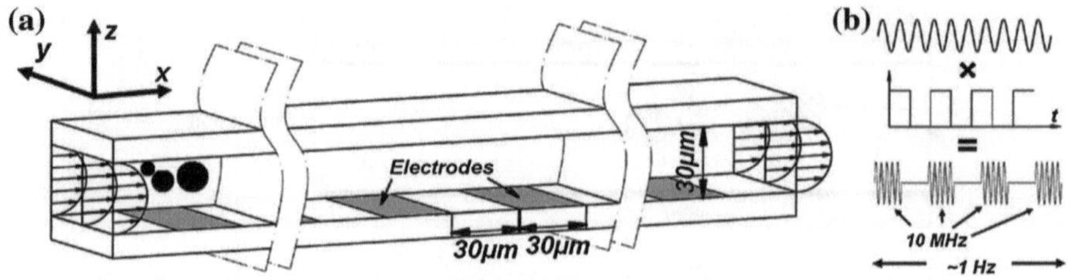

Fig. 10 **a** Schematic of a pulsed-DEP device employing a discontinuous (in time) nDEP force for size-dependent separation of particles [86]. **b** The pulsed-DEP waveform is obtained by time-multiplexing a sine wave with a square wave

The pulsed-DEP waveform is formed by time-multiplexing a sine wave with a square wave as illustrated in Fig. 10b. The analysis of pulsing frequency versus applied voltage provides a recommended operating condition for retaining or releasing any particle size (3, 5 and 10 µm polystyrene beads). The technique was also utilized in combination with a silicon nanotweezer to stretch and trap a single DNA molecule with 50 ms pulses of an 1 MHz AC voltage [87]. The use of pulsed DEP is more advantageous than the conventional AC DEP because when an electric field is pulsing, the electric field-induced flow is lessened, giving rise to the DNA trapping in a more stable environment, and continuous trapping of DNA molecules at the tip of the nanotweezer can be avoided.

3.1.4 Optically-Induced Dielectrophoresis

While many studies focused on DEP techniques with metallic planar electrodes, a new optically-induced DEP technique was developed [88]. As shown in Fig. 11, this technique mainly replies on virtual DEP electrodes made from a photosensitive surface sandwiched by indium tin oxide (ITO)-coated or conductive glass substrates was integrated for the first time. When the photosensitive areas are projected with light, the impedance of projected areas is reduced, thus yielding virtual electrodes for DEP manipulation. The projection of optical images (e.g., circular or parallel electrode patterns) on the photosensitive areas is achieved by combining a light source from a light-emitting diode and a digital micromirror display (DMD) through a 10× objective lens. The photosensitive surface consists of three thin layers: a 50 nm n + hydrogenated amorphous silicon (a-Si:H) layer, an 1 µm undoped a-Si:H layer, and a 20 nm silicon nitride layer. An AC electric field is imposed on both top and bottom ITO-coated electrodes. The intensity of light sufficient to activate virtual electrodes is found to be 100,000 times lower than that used in the laser-based technique such as optical tweezers [89, 90]. Thus, the technique clearly becomes a benevolent tool for cellular applications. This technique is named as optoelectronic tweezers (OET). Because of the flexibility of the OET platform in defining the light-induced electrode patterns, many research groups have proposed a variety of OET systems that can produce dynamic patterns for discriminating particles by size [91], switchable patterns for particle sorting

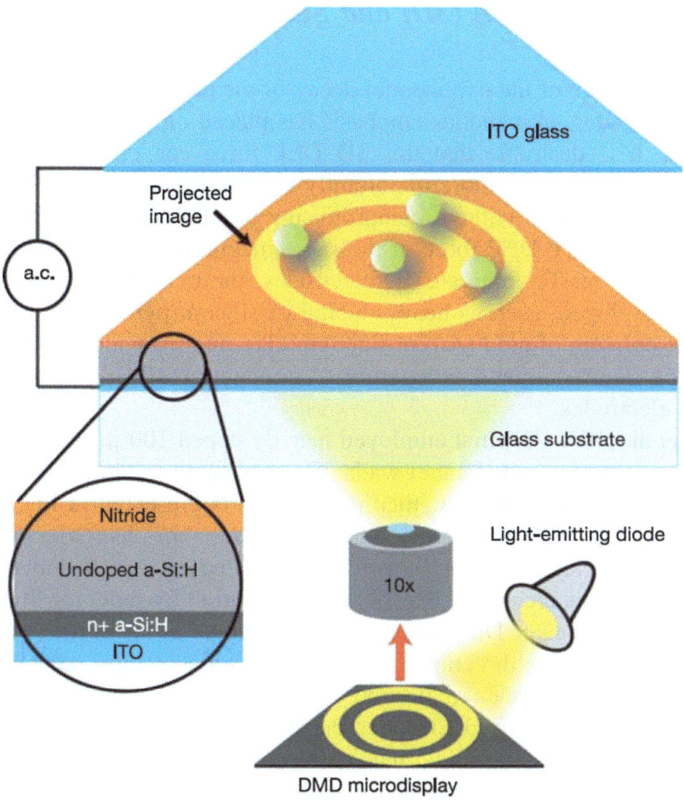

Fig. 11 An optoelectronic tweezer system for DEP manipulation of particles and cells via optically-induced photosensitive electrodes [88]

[92], gray-scale patterns for cell sorting and separation [93, 94], and configurable approaches for cell patterning [95]. Besides lasers as light source, liquid crystal display (LCD) projector [96], len-less direct image [97], LCD integrated with condenser lens [98] were also integrated in the OET systems. In the literature, this OET technology has been continually advanced and employed to manipulate various biosamples such as human B cells (lymphocytes) [88], DNA [99] and CNTs [100], while its applications can be found not only in massively parallel manipulation of cells but also in high-resolution manipulation of single cells [88].

However, like planar electrodes, the OET technique still suffers from an exponential decay of electric fields from the photosensitive surface. Compared to the use of metallic electrodes, the use of ITO-coated glass substrates may also lead to the need of high voltage supply as the light-intensity-dependent impedance of the photosensitive layer is not very sensitive. Furthermore, for fluorescence-based applications, the use of high intensity of light is not always favorable since the differentiation of cells stained by fluorescent dyes may encounter difficulties due to photobleaching.

3.2 Three-Dimensional (3D) and Sidewall Electrodes

To address the issue of the exponential decay of the electric field from the surface of planar electrodes, tremendous emphasis has placed on creating DEP force in a 3D manner. It is desirable that this 3D DEP force can produce strong lateral separation of samples and more importantly can be present uniformly throughout the microchannel height without compromising the throughput. Then cells at any height can experience DEP force of the same magnitude, thereby allowing more efficient separation. To fabricate 3D electrodes, the main fabrication techniques have been developed, such as heavy doping of silicon, pyrolysis of SU-8 photoresist, electroplating of gold (Au) or titanium (Ti), and manual insertion of etched copper (Cu) electrodes, and integration of conducting silver (Ag)-filled PDMS composite electrodes.

Iliescu et al. [101–104] first employed heavily doped 100 μm-thick silicon (Si) as both 3D electrodes and channel walls. The advanced version of Si-based DEP devices with asymmetric doped silicon electrodes is presented in Fig. 12a [105]. During separation, pDEP cells are attracted to the thin electrodes (1 μm-thick amorphous silicon layer) and nDEP cells are levitated and repelled away from the thin electrodes. This technique benefits cell separation by reducing Joule heating and producing a stronger DEP force nearly two times in the vertical direction as compared to the planar direction [105, 106]. In spite of these advantages, it is difficult for users to obtain a desired channel thickness because the channel height depends on the doped Si wafer thickness which is usually much thicker than typical range of microfluidic height. In the meantime, Park et al. presented carbon electrodes for flow-through DEP separation and filtration of cells. With the carbon microelectromechanical system (C-MEMS) technologies, patterned SU-8

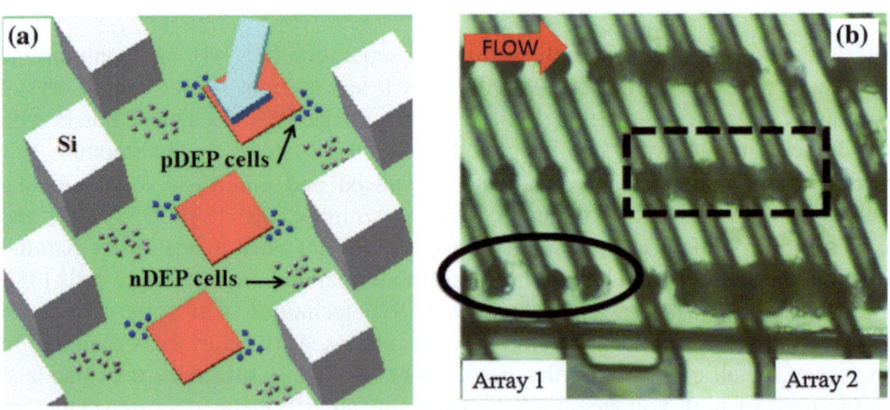

Fig. 12 3D electrodes for continuous particle separation. **a** Heavily doped asymmetric silicon electrodes attracting pDEP cells to the thin electrodes while levitating and repelling nDEP cells away from the thin electrodes [105]. **b** Two arrays of carbon electrodes capturing viable yeast cells at array 1 (*ellipse*) and non-viable yeast cells at array 2 (*rectangle*) [108]

photoresist microstructures can be turned to 3D carbon electrodes through pyrolysis usually above 900 °C in an inert environment [107]. As shown in Fig. 12b, dense arrays of high aspect ratio of carbon electrodes as multi-stage filters (electrically independent) were fabricated and demonstrated to fractionate viable from non-viable yeast cells with a flow rate of 10 μl/min [108]. Numerical analyses of the geometrical effects of carbon electrodes on electric field distributions provided an optimal design [109].

Alternatively, 3D DEP electrodes can be embedded in a sidewall of microchannels to achieve lateral separation of biosamples in continuous flow. Figure 13a depicts a schematic of a DEP device with electroplated Au sidewall electrodes which produce dual-frequency DEP forces causing distinct lateral locations of subpopulations [110]. As this technique controls the channel height by photoresist-coating process, adjustment of the channel height becomes flexible and plausible for high throughput [111]. The device was successfully used to achieve focusing, sorting and separation of beads and biological cells such as mouse neural stem cells and human kidney cells [110–112]. Similarly, sidewall Ti electrodes was also reported for particle manipulation [113]. Using these electroplated sidewall electrodes, the problem of rapid decay of electric fields along the channel height can be improved and the adjustment of channel height is flexible. However, these microfluidic devices are usually made from glass or silicon, and prevention of liquid leakage becomes a problem [111].

With rapid development of soft lithography, polydimethylsiloxane (PDMS) is now widely used as a fabrication material in microfluidic devices, owing to beneficial features: ease of fabrication and bonding, optical transparency, biocompatibility, and low cost [9, 10, 116]. However, it remains difficult to integrate 3D or sidewall metal electrodes for producing AC-DEP in PDMS-based microfluidic devices, because adhesion between metallic electrodes and PDMS is poor [117]. To date, two fabrication techniques have been reported: use of a manual insertion of etched Cu electrodes [114, 118] and fabrication of 3D conducting AgPDMS composite electrodes in a PDMS microchannel [115, 119].

Cetin et al. [118] and Kang et al. [114] reported a manual insertion of etched Cu sheets as sidewall electrodes in PDMS-based devices. The difference of these two studies is that Cetin et al. utilized the sidewall Cu electrodes embedded in a straight channel while Kang et al. [114] introduced a hybrid design using a PDMS hurdle and sidewall Cu electrodes as shown in Fig. 13b. As DEP forces occur only near the corners of the PDMS hurdle, this leads to reduced exposure of electric field exerting on cells and also mitigates negative effects related to Joule heating. Such hybrid-based device was tested to separate microparticles from yeast cells. Even though the device assembly between a glass cover and PDMS channels using oxygen plasma treatment is simple, this device fabrication method may not be effective and convenient because of the following reasons. First, the etching process for patterning Cu electrodes cannot guarantee the exact dimension of electrodes to match with PDMS channels, and thus there exists gaps between the inserted electrodes and PDMS walls. Such gaps can cause fluid leakage and even cell loss during separation (Fig. 13b). Second, similar to the case of heavily doped

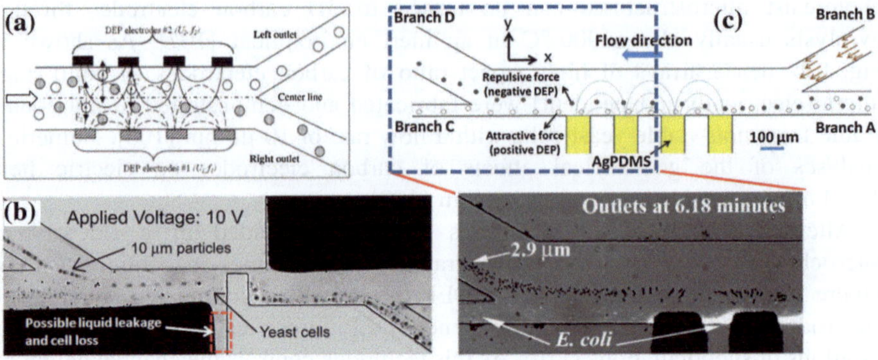

Fig. 13 Sidewall electrodes for continuous-flow separation of cells. **a** Electroplated Au sidewall electrodes with dual-frequency DEP forces resulting in distinct lateral locations of subpopulations [110]. **b** Manually inserted Cu electrodes for separating 10 μm particles by nDEP from yeast cells [114]. **c** Sidewall conducting AgPDMS electrodes for separating *E. coli* from 2.9 μm particles of similar size [115]

silicon, adjustment of channel height is constrained by the availability of Cu sheet thickness. Third, it becomes difficult for batch fabrication when manual insertion of multiple Cu electrodes is necessary to enhance DEP force for higher throughput.

Our group recently developed fabrication technique using the synthesis and the integration of 3D sidewall conducting AgPDMS composite electrodes in PDMS microchannels [115, 119, 120]. As depicted in Fig. 13c, sidewall AgPDMS electrodes are able to attract a tremendous number of *E. coli* and to repel 2.9 μm particles of similar size. The separation efficiency of 97 % was achieved. The devices also demonstrated separation of 5, 10 and 15 μm particles by size [119] and separation of yeast cells from 5 μm particles of similar sizes by polarizability [115]. In addition to the continuous separation, DEP characterization of submicron, microparticles, yeast and bacterial cells were successfully demonstrated [115, 120]. As the AgPDMS composites are synthesized by dispersing 1 μm silver fillers into PDMS gel (85 % w/w), the PDMS cohesion strengthens the bond between the conducting AgPDMS composites and the PDMS channel. Thus, this approach greatly facilitates (1) the integration of the conducting AgPDMS electrodes with PDMS microstructures; and (2) the assembly of the device through bonding between the PDMS channels, with embedded PDMS composites electrodes, and a glass substrate, involving only oxygen plasma treatment. Moreover, since the electrode height is controlled by the photoresist thickness, selection of desirable channel height and integration of multiple sidewall electrodes are flexible and practically possible for batch fabrication. As compared to other 3D or sidewall electrode-based devices, this device with AgPDMS electrodes is deemed the first and important step towards a sole polymer-based microfluidic platform offering multiple functions (e.g., focusing, sorting, separation and characterization of particles and cells).

3.3 Insulator-Based Dielectrophoresis (iDEP)

Instead of utilizing metallic electrodes, Cummings and Singh proposed a new class of DEP devices which consist of an array of electrically insulating posts as shown in Fig. 14a for generating non-uniform electric field [121]. Usually, a DC voltage is imposed via external electrodes placed at the inlet and the outlet. Due to the change of electrical current density around the insulating posts, spatial gradients of electric field can be induced. This insulator-based DEP (or iDEP) offers relatively low costs and is suitable for mass fabrication through the use of injection molding or hot embossing, and suffers less severe electrolysis and fouling problems because of no electrodes fabricated inside channels [122, 123]. In addition, similar to 3D or sidewall electrodes, iDEP can also create 3D DEP forces across the channel height.

The iDEP principle of sample transport is originally based on the DC-driven coupled phenomena, including linear electrophoresis (EP) and electroosmosis (EOF) and nonlinear DEP. EP refers to the motion of an electrically charged sample in a stationary electrolyte solution under an applied electric field while EOF refers to the fluid motion due to the interaction between the applied electric field and electric double layer (EDL) formed next to the charged wall of a microchannel. Generally, EP and EOF occur simultaneously because both the sample and the channel surface are charged. As a result, in iDEP devices, EP and EOF are for transporting samples in a microchannel and DEP force is for separating or trapping the samples.

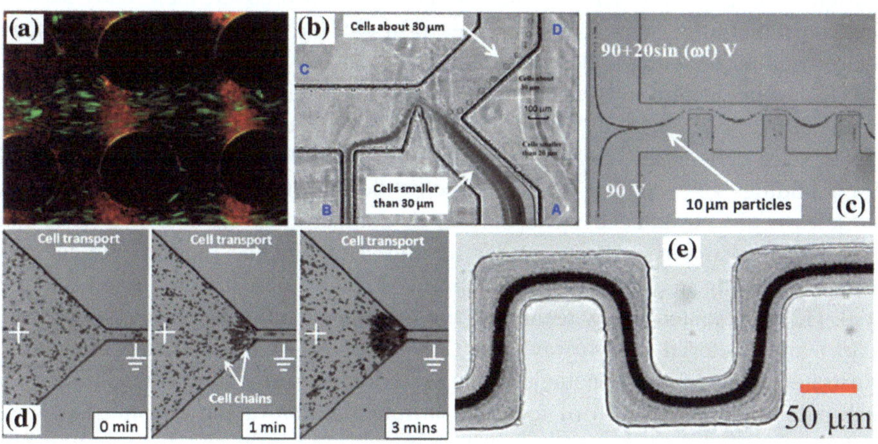

Fig. 14 Use of iDEP for particle and cell manipulations. **a** Glass-etched iDEP chip containing insulating circular posts for DC-based separation of *B. subtilis* (*labeled green*-Syto® 11) and *B. cereus* (*labeled red*-Syto® 17) [121]. **b** DC-induced lateral separation of small and large breast cancer cells by using a triangular PDMS hurdle [124]. Applications of DC-offset AC electric field for: **c** sorting of 10 μm particles using multi-insulating blocks at 2 kHz [125], **d** DEP concentration of yeast cells at 5 kHz [126] and **e** separation of focused yeast cells (*black*) and unfocused bacterial cells (*gray*) via DEP focusing in a serpentine channel at 1 kHz [127]

Figure 14a presents the separation of *B. subtilis* (labeled green-Syto® 11) and *B. cereus* (labeled red-Syto® 17) in a glass-etched iDEP chip containing arrays of insulating circular (quartz) posts [121]. This DC-iDEP principle has been successfully demonstrated for both separating and concentrating submicron particles, DNA, bacteria, protein and blood cells by using various insulator structures such as triangular and sawtooth constrictions and faceted prism microchannels [121, 128–132]. In addition, lateral DC-DEP separation of particles and cells was achieved by using an oil meniscus [133] and PDMS blocks [125, 134]. A serpentine channel was demonstrated for focusing microparticles [135]. Another example is lateral separation of small and large breast cancer cells by using a triangular PDMS hurdle as depicted in Fig. 14b [124]. However, the drawback of the DC-iDEP technique is the requirement of high DC voltages due to the aforementioned coupled linear EP and EOF and nonlinear DEP electrokinetic phenomena [123, 126, 136], and the undesirable consequence leads to Joule heating affecting viability and physiology of the cells [137].

Other than the use of sole DC field, a number of studies have been reported that the use of a DC-offset AC electric field is promising for manipulating particles and cells in iDEP devices. Hawkins et al. [27] employed the DEP force induced by a DC-offset AC electric field to sort and trap particles in a microchannel with 3D constriction along the depth. Their study showed an important finding that with an applied DC-offset AC field, EOF and EP force could be decoupled from DEP force, thus providing independent control of these electrokinetic forces. Lewpiriyawong et al. [125] further discovered that the use of DC-offset AC field is capable of minimizing Joule heating by reducing the total voltage required for particle sorting (as compared to the sole DC field). Moreover, their study showed that the iDEP effect can focus particles in a similar fashion as the well-known hydrodynamic focusing and thus facilitate the particle sorting and separation in a PDMS H-filter with multi-insulating blocks as shown in Fig. 14c. Lewpiriyawong et al. also reported a study using the DC-offset AC electric field to concentrate viable yeast cells in a PDMS channel with a tapered contraction (Fig. 14d) [126]. Their findings showed that the total electric field required to concentrate 15 μm particles could be reduced up to 85.9 % as compared to the sole DC field. Zhu et al. [138] also reported the similar findings showing the reduced voltages for particle focusing in a constriction-integrated PDMS microchannel. Later, Church et al. [127] presented a new technique for size-dependent separation of yeast cells (dark) and bacterial cells (gray) via DEP focusing in a serpentine channel as illustrated in Fig. 14e. Nonetheless, these studies reported similar results that the manipulation of the width of focused particle streams or the enhanced concentration of particles can be achieved by adjusting the ratio of AC-to-DC electric fields.

Although the effect of Joule heating is much less severe in iDEP devices under the applied DC-offset AC electric field, the Joule heating still remains and can be more significant in biological fluids with high conductivity. Hawkins et al. [139] showed that even though the solution conductivity was $>10^{-3}$ S/m, strong electrothermally induced (ET) flow existed near an insulating block with a 25 μm constriction in a 100 μm microchannel. This ET flow perturbs the local EOF flow field, thus generating fluid circulation (or vorticity). Instead of deteriorating the device performance, their findings showed vorticity-aided separation and trapping of those particles experiencing nDEP. A similar investigation conducted by Sridharan et al. [140] found that counter-rotating fluid circulations can be manipulated to occur at either downstream end alone or both ends of the constriction channel, depending on the applied DC magnitude.

As can be seen from the literature describing from the use of DC electric field to the DC-offset AC electric field, Joule heating causing temperature rise has been a main issue. High temperature not only may kill or lyse the processed samples but also influence the local EOF flow field. Therefore, it is of an importance to consider the effect of Joule heating and the ET flow as crucial factors when designing iDEP devices to deal with biosamples in physiological fluids typically with high conductivity (~0.1–1 S/m) [141].

3.4 Hybrid DEP

Hybrid DEP combines both planar-electrode-based and insulator-based DEP techniques for continuous separation and trapping of biosamples [142, 143]. The key features of a hybrid DEP system then are (1) the ability to manipulate samples with low voltages like planar DEP systems and (2) straightforward mass fabrication by rapid prototyping like iDEP systems [123]. So far, two types of the hybrid DEP have been proposed: (1) use of liquid electrodes and (2) use of contactless DEP. Figure 15a illustrates a schematic of a liquid electrode DEP system in which the comb-liked SU-8 microstructures are fabricated side by side with Ti/Pt planar electrodes [142]. An electric field is produced by planar electrodes (in black) in contact with the flowing fluid and the patterned SU-8 insulator is responsible for creating nonuniform electric field. Using multiple electric field frequencies, the downstream equilibrium locations of subpopulations can be controlled by DEP forces created near insulating structures at both sides of the microchannel. The technique has been successfully used for focusing and continuous separation of particles, live/dead yeast cells, *Babesia bovis*-infected red blood cells and platelets [142, 144–148]. Details of the optimized design of liquid electrodes can be found elsewhere [149]. In spite of the advantages and high performance of the liquid electrode-based devices, the use of SU-8 photoresist as channel walls can be expensive especially for disposable or point-of-care devices. Also, fouling can be an issue if the device is operated with real biosamples (e.g., blood) for a long period of time. The use of inexpensive materials like PDMS or other polymers can be an option to make this technique more promising and reliable for clinical uses.

The second hybrid DEP system is named as contactless DEP (cDEP) because it produces a nonuniform electric field without direct contact between metal electrodes and the sample fluid (Fig. 15b) [143]. This new system clearly can

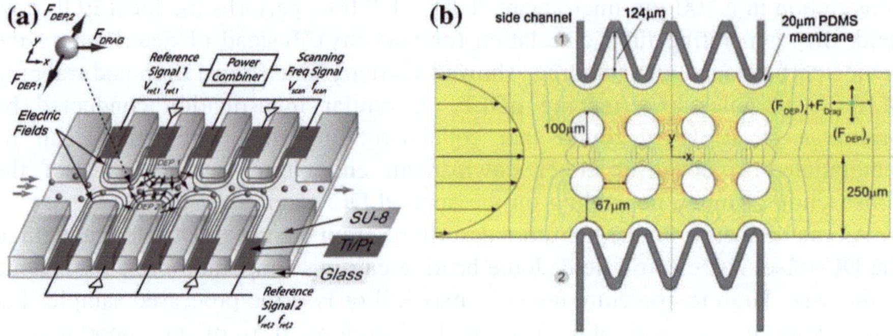

Fig. 15 Two hybrid DEP systems: **a** schematic of a flow-based DEP system with liquid electrodes comprising the comb-liked SU-8 microstructures fabricated side by side with Ti/Pt planar electrodes [142]. **b** Contactless DEP device with 100 μm circular insulating posts used as flow-based DEP filtering system where nonuniform electric field is generated from the zigzag microchannel filled with a highly conductive medium separated by a thin PDMS membrane [143]

avoid bubble generation and reduce severe fouling and any possible contamination during separation. To generate electric field in the main channel, an AC voltage is applied via external aluminium electrodes remotely placed at the zigzag microchannel reservoirs. Because these zigzag microchannels contain a highly conductive medium, although separated from the main channel by a thin insulating PDMS layers, a nonuniform electric field and thus the DEP force can be created near the circular posts without directly immersing metal electrodes in the sample fluid. Phosphate buffered saline (PBS) was chosen as the conductive medium filled in the zigzag channels. Such cDEP devices have been demonstrated to successfully separate various biosamples such as live and dead cells [143] and human cancer cells [150]. In addition to these two hybrid DEP designs, researchers also proposed similar hybrid DEP devices by employing other insulating structures such as a micropillar array to discriminate or trap particles by size [151] and inclined rod-liked insulators to focus and filter biological cells [152].

4 Concluding Remarks

This chapter presents an overview of the development of DEP-FFF technologies for continuous-flow separation of biosamples in microfluidics. The publications between 2005 and 2012 are the main focus of this report. Based on the strategies in providing nonuniform electric field for DEP force, DEP devices are categorized into four groups: (1) the use of planar (or thin film) electrodes, (2) the use of 3D and sidewall electrodes, (3) the use of insulating topographical structures, and finally (4) the hybrid DEP systems. Despite a plethora of proposed DEP devices and their successes, the efficiency, throughput and selectivity of these reported DEP devices still do not meet the requirements of clinical laboratories and cannot be competed with the existing commercial biomedical analytical instrument such as flow cytometry and FACS.

The capability of performing continuous separation of biosamples with high efficiency and throughput is highly desirable in DEP microfluidic devices. Currently, most DEP devices can be operated at about 1–10 μl/min while the working flow rate of clinical and research facilities such as FACS (LSR II cell analyzer, BD Biosciences, San Jose, California) is about 60 μl/min [72, 153]. In principle, the high throughput of DEP systems can readily be achieved by increasing both the flow rate and the applied voltages. However, this is often not practical because upon increasing voltages, current-induced Joule heating rapidly causes temperature rise in the DEP systems (including both electrode-based and insulator-based systems). As a result, the sample can be potentially killed in high temperature environment (e.g., >41 °C) above its physiological temperature even for a short period of time [154]. In addition, bubble generation induced by highly applied voltages can interrupt separation process. One possible solution is to integrate a cooling system near the heat-generating area of electrodes. Recently, Moschallski et al. [46] demonstrated the feasibility of removing heat by attaching a chilled water system to the bottom face of the DEP device. With such cooling system, the device could be operated at 4–10 °C which can preserve integrity of organelle during the 60 h continuous-flow separation. In addition, a Peltier cooling pad was also used to maintain the desired temperature in a DEP device [155]. Therefore, it is expected to further develop DEP devices with more efficient cooling systems such that higher DEP system throughput can be achieved.

Nanoscale electrodes, typically fabricated via electron beam (e-beam) lithography, may play an important role in future DEP technologies by increasing throughput, efficiency and selectivity in separating rare cells in continuous flow. It has been shown since 1996 that the e-beam lithography allows for the fabrication of electrodes with dimensions from a few micron to 60 nm with ZEP-520 resist [156] or even 20 nm with PMMA resist [157]. In comparison with microscale electrodes, when nanoscale electrodes are energized under similar magnitude of AC voltages, DEP force produced by the nanoscale electrodes can be dramatically increased. Hence, it is anticipated that DEP devices can achieve separation with higher efficiency and also require lesser energy consumption. Moreover, as the electrode area significantly reduces, the effect of Joule heat is also greatly minimized due to the high surface-to-volume ratio [158]. However, in the past five years, DEP devices still relied heavily on microscale electrodes with dimensions of 50–100 μm. Only a limited number of studies were reported to employ microscale electrodes with relatively small dimensions, about 2 μm using e-beam lithography [159] and 10 μm using photolithography [52, 160]. The device functions were manifested to trap 282 nm particles [159], focus 40 nm particles at 2 mm/s [52] and sort 6 μm particles at 10 mm/s [160] under an AC voltage of 15–20 V peak-to-peak. To date, no study has been to utilize such nanoscale electrodes for continuous-flow separation of biological cells. In the near future, the implementation of nanoscale electrodes could be a promising and challenging research direction to realize fast separation and detection of pathogenic rare cells, virus, protein and

DNA with high efficiency, throughput and selectivity. It should be pointed out here that nanoscale insulating posts can also be used to induce highly nonuniform electric field in iDEP devices for enhanced separation efficiency.

Another important ongoing trend is an attempt to increase sensitivity of DEP biosensors for foodborne and waterborne pathogens detection. The presence of pathogens in food and water often results in a deathly impact on many groups of people: the elderly, pregnant women, newborns and even healthy persons [161]. Among pathogenic microorganisms, *L. monocytogenes*, *Listeria*, *Salmonella*, *E. coli* O157:H7 *and* protozoan (Cryptosporidium) are the leading causes of foodborne and waterborne lethal illness [162–164]. For safe drinking water, the concentration of *E. coli* O157:H7 must be lower than 1 Colony Forming Units (CFU)/ml [165] and around 1–30 CFU/ml for clear clinical symptoms to manifest [166]. Due to such a low concentration, pathogenic detection becomes rather difficult in practical. The plate-culture method is a traditional detection approach (still widely used nowadays) which is tedious, time consuming (a few days), and labour costs (as it requires a well trained personnel to identify bacterial colony morphology to provide the readings) [167]. To date, the detection limit for currently available sensors [e.g., enzyme linked immunosorbent assay (ELISA) and polymerase chain reaction (PCR)] are on the order of about 10^1–10^6 CFU/ml [168]. Clearly, currently available sensors are still not sensitive enough for detecting those pathogens. However, it has been reported in the DEP community that the utilization of antibodies is promising in improving the selectivity of DEP systems where antibodies can typically be either immobilized on the DEP sensor surface or coated on flowing micro/nanoparticle surface [58, 61, 169]. Yang et al. [170] showed that the DEP-assisted capture efficiency of *Salmonella* cells by immobilized Anti-*Salmonella* antibodies on the DEP chip with planar electrodes was significantly improved from 17.6 to 64 % in a non-continuous separation mode. In a continuous separation mode, Kim et al. [58] achieved separation of multiple bacterial cells with distinct synthetic DEP tags in a DEP device with inclined double-sided electrodes. Their device demonstrated 1,000-folder enrichment of multiple bacterial cells at a throughput of $\sim$4,166 particles/min in a single pass separation. Similar development was the use of magnetic nanoparticles-antibody conjugates (MNAC) to assist in detecting *E. coli* O157:H7 at concentrations as low as 1.6×10^2 and 1.2×10^3 (numbers of bacteria in 60 nL loaded in a PDMS microchannel). These concentrations are equivalent to 94 and 74 % in capture rates for pure culture and ground beef samples, respectively [164]. Besides planar electrodes, a 25 μm microwire was recently demonstrated for detecting *E. coli* K-12 in baby spinach leaves with the detection limit of 10^3 CFU/ml [165]. Although improved, such detection limit was merely the pathogenic concentration of patients near death [167]. Therefore, the early pathogenic detection is still challenging. Based on the literature, only planar electrodes have been employed in DEP biosensors while other kinds of DEP devices such as 3D/sidewall electrodes, insulating structures or hybrid DEP systems could be used as new biosensing platforms. Last but not least, as mentioned earlier, the cooling system and nanoscale electrodes can be important elements in developing efficient and reliable biosensors. For more information on DEP fractionation and detection of other rare cells such as blood cells, cancer cells and mammalian cells, the readers may consult an excellent review by Pratt et al. [171].

Acknowledgments The authors would acknowledge the research grant (AcRF RG17/05) from the Ministry of Education of Singapore to CY and the Ph.D. Scholarship from Nanyang Technological University to NL. The authors also are grateful to Dr. Marcos for his comments and critical reading of the manuscript.

References

1. Toner, M., Irimia, D.: Blood on a chip. Annu. Rev. Biomed. Eng. **7**, 77–103 (2005)
2. Cho, B.S., et al.: Passively driven integrated microfluidic system for separation of motile sperm. Anal. Chem. **75**, 1671–1675 (2003)
3. Greenberg, A.E., Clesceri, L.S., Eaton, A.D.: *Standard Methods for the Examination of Water and Wastewater*, 21st edn. American Public Health Association, American Water Works Association, and Water Environment Federation (2005)
4. Lenshof, A., Laurell, T.: Continuous separation of cells and particles in microfluidic systems. Chem. Soc. Rev. **39**, 1203–1217 (2010)
5. Dainiak, M.B., et al.: Cell chromatography: Separation of different microbial cells using IMAC supermacroporous monolithic columns. Biotechnol. Prog. **21**, 644–649 (2005)
6. Hulett, H.R., et al.: Cell sorting: automated separation of mammalian cells as a function of intracellular fluorescence. Science **166**, 747–749 (1969)
7. Yager, P., et al.: Microfluidic diagnostic technologies for global public health. Nature **442**, 412–418 (2006)
8. Bonner, W.A., et al.: Fluorescence activated cell sorting. Rev. Sci. Instrum. **43**, 404–409 (1972)
9. El-Ali, J., Sorger, P.K., Jensen, K.F.: Cells on chips. Nature **442**, 403–411 (2006)
10. Bilitewski, U., et al.: Biochemical analysis with microfluidic systems. Anal. Bioanal. Chem. **377**, 556–569 (2003)
11. Schimpf, M.E., Caldwell, K., Giddings, J.C.: Field-Flow Fractionation handbook. Principle and Theory. John Wiley & Sons, Inc., New York (2000)
12. Reschiglian, P., et al.: Field-flow fractionation and biotechnology. Trends Biotechnol. **23**(9), 475–483 (2005)
13. Huang, Y., et al.: Introducing dielectrophoresis as a new force field for field-flow fractionation. Biophys. J. **73**, 1118–1129 (1997)
14. Yang, J., et al.: Cell separation on microfabricated electrodes using dielectrophoretic/gravitational field-flow fractionation. Anal. Chem. **71**, 911–918 (1999)
15. Wang, X.-B., et al.: Cell separation by dielectrophoretic field-flow fractionation. Acc. Chem. Res. **72**, 832–839 (2000)
16. Song, H., et al.: Continuous-mode dielectrophoretic gating for highly efficient separation of analytes in surface micromachined microfluidic devices. J. Micromech. Microeng. **18**, 125013 (2008)
17. Markx, G.H., Talary, M.S., Pethig, R.: Separation of viable and non-viable yeast using dielectrophoresis. J. Biotechnol. **32**, 29–37 (1994)
18. Nedelcu, S., Watson, J.H.P.: Size separation of DNA molecules by pulsed electric field dielectrophoresis. J. Phys. D Appl. Phys. **37**, 2197–2204 (2004)
19. Pohl, H.A.: The motion and participation of suspensoids in divergent electric fields. J. Appl. Phys. **22**(7), 869–871 (1951)

20. Pohl, H.A.: Dielectrophoresis: The Behavior of Neutral Matter in Nonuniform Electric Fields. Cambridge University Press, Cambridge (1978)
21. Jones, T.B.: Electromechanics of Particles. Cambridge University Press, Cambridge (1995)
22. Lewpiriyawong, N.: Continuous Separation and Manipulation of Particles and Cells Using Dielectrophoresis. Nanyang Technological University, Singapore (2011)
23. Wang, X., Wang, X.-B., Gascoyne, P.R.C.: General expressions for dielectrophoretic force and electrorotational torque derived using the Maxwell stress tensor method. J. Electrostat. **39**, 277–295 (1997)
24. Denner, V., Pohl, H.A.: Dielectrophoretic force in electrostatic fields. J. Electrostat. **13**, 167–174 (1982)
25. Morgan, H., Green, N.G.: AC Electrokinetics: Colloids and Nanoparticles. Research Studies Press, Philadelphia (2002)
26. Arnold, W.M., Schwan, H.P., Zimmermann, U.: Surface conductance and other properties of latex particles measured by electrorotation. J. Phys. Chem. **91**, 5093–5098 (1987)
27. Hawkins, B.G., et al.: Continuous-flow particle separation by 3D insulative dielectrophoresis using coherently shaped, dc-biased, ac electric fields. Anal. Chem. **79**, 7291–7300 (2007)
28. Wang, X.-B., et al.: A unified theory of dielectrophoresis and travelling wave dielectrophoresis. J. Phys. D Appl. Phys. **27**, 1571–1574 (1994)
29. Nguyen, N.T., Wereley, S.: Fundamentals and Applications of Microfluidics, 2nd edn. Artech House, Boston (2006)
30. Wang, X.-B., et al.: Separation of polystyrene microbeads using dielectrophoretic/gravitational field-flow-fractionation. Biophys. J. **74**, 2689–2701 (1998)
31. Gascoyne, P., Satayavivad, J., Ruchirawat, M.: Microfluidic approaches to malaria detection. Acta Trop. **89**, 357–369 (2004)
32. Markx, G.H., Pethig, R., Rousselet, J.: The dielectrophoretic levitation of latex beads with reference to field-flow fractionation. J. Phys. D Appl. Phys. **30**, 2470–2477 (1997)
33. Markx, G.H., Rousselet, J., Pethig, R.: DEP-FFF: Field-flow fractionation using non-uniform electric field. J. Liq. Chromatogr. Relat. Technol. **20**(16-17), 2857–2872 (1997)
34. Rousselet, J., Markx, G.H., Pethig, R.: Separation of erythrocytes and latex beads by dielectrophoretic levitation and hyperlayer field-flow fractionation. Colloids Surf., A **140**, 209–216 (1998)
35. Holmes, D., Green, N.G., Morgan, H.: Microdevices for dielectrophoretic flow-through cell separation. IEEE Eng. Med. Biol. Mag. **22**(6), 85–90 (2003)
36. Markx, G.H., Pethig, R.: Dielectrophoretic separation of cells: Continuous separation. Biotechnol. Bioeng. **45**, 337–343 (1995)
37. Ino, K., et al.: Manipulation of microparticles for construction of array patterns by negative dielectrophoresis using multilayered array and grid electrodes. Biotechnol. Bioeng. **104**(4), 709–718 (2009)
38. Thomas, R.S.W., et al.: Trapping single human osteoblast-like cells from a heterogeneous population using a dielectrophoretic microfluidic device. Biomicrofluidics **4**, 022806 (2010)
39. Jang, L.-S., Huang, P.-H., Lan, K.-C.: Single-cell trapping utilizing negative dielectrophoretic quadrupole and microwell electrodes. Biosens. Bioelectron. **24**, 3637–3644 (2009)
40. Ramadan, Q., et al.: Simultaneous cell lysis and bead trapping in a continuous flow microfluidic device. Sens Actuators B **113**, 944–955 (2006)
41. Auerswald, J., et al.: Fast immobilization of probe beads by dielectrophoresis-controlled adhesion in a versatile microfluidic platform for affinity assay. Electrophoresis **26**, 3697–3705 (2005)
42. Lagally, E.T., Lee, S.-H., Soh, H.T.: Integrated microsystem for dielectrophoretic cell concentration and genetic detection. Lab Chip **5**, 1053–1058 (2005)
43. Gascoyne, P.R.C.: Dielectrophoretic-field flow fractionation analysis of dielectric, density, and deformability characteristics of cells and particles. Anal. Chem. **81**, 8878–8885 (2009)

44. Morgan, H., et al.: The dielectrophoretic and travelling wave forces generated by interdigitated electrode arrays: Analytical solution using Fourier series. J. Phys. D Appl. Phys. **34**, 1553–1561 (2001)
45. Durr, M., et al.: Microdevices for manipulation and accumulation of micro- and nanoparticles by dielectrophoresis. Electrophoresis **24**, 722–731 (2003)
46. Moschallski, M., et al.: MicroPrep: Chip-based dielectrophoretic purification of mitochondria. Electrophoresis **31**, 2655–2663 (2010)
47. Chen, D., Du, H.: A dielectrophoretic barrier-based microsystem for separation of microparticles. Microfluid. Nanofluid. **3**, 603–610 (2007)
48. Chen, D.F., Du, H., Li, W.H.: A 3D paired microelectrode array for accumulation and separation of microparticles. J. Micromech. Microeng. **16**, 1162–1169 (2006)
49. Choi, S., Park, J.-K.: Microfluidic system for dielectrophoretic separation based on a trapezoidal electrode array. Lab Chip **5**, 1161–1167 (2005)
50. Yang, F., et al.: Dielectrophoretic separation of colorectal cancer cells. Biomicrofluidics **4**, 013204 (2010)
51. Khoshmanesh, K., et al.: Dielectrophoretic-activated cell sorter based on curved microelectrodes. Microfluid. Nanofluid. **9**, 411–426 (2010)
52. Morgan, H., Holmes, D., Green, N.G.: 3D focusing of nanoparticles in microfluidic channels. IEE Proc.-Nanobiotechnol. **150**(2), 76–81 (2003)
53. Ling, S.H., Lam, Y.C., Kua, C.H.: Particle streaming and separation using dielectrophoresis through discrete periodic microelectrode array. Microfluid. Nanofluid. **11**(5), 579–591 (2011)
54. Park, J., et al.: An efficient cell separation system using 3D-asymmetric microelectrodes. Lab Chip **5**, 1264–1270 (2005)
55. Liao, S.-H., Cheng, I.-F., Chang, H.-C.: Precisely sized separation of multiple particles based on the dielectrophoresis gradient in the z-direction. Microfluid. Nanofluid. **12**, 201–211 (2011)
56. Yang, F., et al.: Cascade and staggered dielectrophoretic cell sorters. Electrophoresis **32**, 2377–2384 (2011)
57. Han, K.-H., Han, S.-I., Frazier, A.B.: Lateral displacement as a function of particle size using a piecewise curved planar interdigitated electrode array. Lab Chip **9**, 2958–2964 (2009)
58. Kim, U., et al.: Multitarget dielectrophoresis activated cell sorter. Anal. Chem. **80**, 8656–8661 (2008)
59. Doh, I., Cho, Y.-H.: A continuous cell separation chip using hydrodynamic dielectrophoresis (DEP) process. Sens. Actuators **121**, 59–65 (2005)
60. Kralj, J.G., et al.: Continuous dielectrophoresis size-based particle sorting. Anal. Chem. **78**, 5019–5025 (2006)
61. Kim, U., Soh, H.T.: Simultaneous sorting of multiple bacterial targets using integrated dielectrophoretic–magnetic activated cell sorter. Lab Chip **9**, 2313–2318 (2009)
62. Pommer, M.S., et al.: Dielectrophoretic separation of platelets from diluted whole blood in microfluidic channels. Electrophoresis **29**, 1213–1218 (2008)
63. Talary, M.S., et al.: Electromanipulation and separation of cells using travelling electric fields. J. Phys. D Appl. Phys. **29**, 2198–2203 (1996)
64. Hughes, M.P., Pethig, R., Wang, X.-B.: Dielectrophoretic forces on particles in travelling electric fields. J. Phys. D Appl. Phys. **29**, 474–482 (1996)
65. Huang, Y., et al.: Electrokinetic behavior of colloidal particles in travelling electric fields: Studies using yeast cells. J. Phys. D Appl. Phys. **26**, 1528–1535 (1993)
66. Gascoyne, P., et al.: Microsample preparation by dielectrophoresis: Isolation of malaria. Lab Chip **2**, 70–75 (2002)
67. Choi, E., Kim, B., Park, J.: High-throughput microparticle separation using gradient traveling wave dielectrophoresis. J. Micromech. Microeng. **19**, 125014 (2009)
68. Morgan, H., et al.: Large-area travelling-wave dielectrophoresis particle separator. J. Micromech. Microeng. **7**, 65–70 (1997)

69. Cui, L., Holmes, D., Morgan, H.: The dielectrophoretic levitation and separation of latex bead in microchips. Electrophoresis **22**, 3893–3901 (2001)
70. Kawamoto, H.: Some techniques on electrostatic separation of particle size utilizing electrostatic traveling-wave field. J. Electrostat. **66**, 220–228 (2008)
71. Huang, C., et al.: Design and fabrication of an automated microchip-based cell separation device. Anal. Lett. **40**, 763–778 (2007)
72. Cheng, I.-F., et al.: A continuous high-throughput bioparticle sorter based on 3D traveling-wave dielectrophoresis. Lab Chip **9**, 3193–3201 (2009)
73. Huang, Y., et al.: Differences in the AC electrodynamics of viable and non-viable yeast cell determined through combined dielectrophoresis and electrorotation studies. Phys. Med. Biol. **37**(7), 1499–1517 (1992)
74. Zhou, X.-F., et al.: Differentiation of viable and non-viable bacterial biofilms using electroortation. Biochim. Biophys. Acta **1245**, 85–93 (1995)
75. Yang, J., et al.: Dielectric properties of human leukocyte subpopulations determined by electrorotation as a cell separation criterion. Biophys. J. **76**, 3307–3314 (1999)
76. Arnold, W.M., Zimmermann, U.: Rotating-field-induced rotation and measurement of the membrane capacitance of single mesophyll cells of Avena sativa. Zeitschrift fuer Naturforschung. Section C. Biosci. **37**, 908–915 (1982)
77. Cristofanilli, M., et al.: Automated electrorotation to reveal dielectric variations related to HER-2/neu overexpression in MCF-7 sublines1. Clin. Cancer Res. **8**, 615–619 (2002)
78. Huang, Y., Wang, X., Becker, F.F., Gascoyne, P.R.: Membrane changes associated with the temperature-sensitive P85gag-mos-dependent transformation of rat kidney cells as determined by dielectrophoresis and electrorotation. Biochim. Biophys. Acta **1282**(1), 76–84 (1996)
79. Pethig, R.J., Sanger, L.M., Heart, R.H., Corson, E., Smith, E.D., Peter, J.S.: Electrokinetic measurements of membrane capacitance and conductance for pancreatic β-cells. IEE Proc.-Nanobiotechnol. **152**(6), 189–193 (2005)
80. Lei, U., Sun, P.-H., Pethig, R.: Refinement of the theory for extracting cell dielectric properties from dielectrophoresis and electrorotation experiments. Biomicrofluidics **5**, 044109 (2011)
81. Schwan, H.P.: Electrical properties of tissue and cell suspensions. Adv. Biol. Med. Phys. **5**, 147–209 (1957)
82. Fuhr, G., Hagedorn, R., Goring, H.: Separation of different cell types by rotating electric fields. Plant Cell Physiol. **26**, 1527–1531 (1985)
83. Kua, C.H., et al.: Dynamic cell fractionation and transportation using moving dielectrophoresis. Anal. Chem. **79**, 6975–6987 (2007)
84. Kua, C.H., et al.: Modeling of dielectrophoretic force for moving dielectrophoresis electrodes. J. Electrostat. **66**, 514–525 (2008)
85. Kua, C.H., et al.: Cell motion model for moving dielectrophoresis. Anal. Chem. **80**, 5454–5461 (2008)
86. Cui, H–.H., et al.: Separation of particles by pulsed dielectrophoresis. Lab Chip **9**, 2306–2312 (2009)
87. Kumemura, M., et al.: Single-DNA-molecule trapping with silicon nanotweezers using pulsed dielectrophoresis. J. Micromech. Microeng. **21**, 054020 (2011)
88. Chiou, P.Y., Ohta, A.T., Wu, M.C.: Massively parallel manipulation of single cells and microparticles using optical images. Nature **436**, 370–372 (2005)
89. Grier, D.G.: A revolution in optical manipulation. Nature **424**, 810–816 (2003)
90. Garces-Chavez, V., Dholakia, K., Spalding, G.C.: Extended-area optically induced organization of microparticles on a surface. Appl. Phys. Lett. **85**, 031106 (2005)
91. Lin, W.-Y., Lin, Y.-H., Lee, G.-B.: Separation of micro-particles utilizing spatial difference of optically induced dielectrophoretic forces. Microfluid. Nanofluid. **8**, 217–229 (2010)
92. Lin, Y.-H., Lee, G.-B.: Optically induced flow cytometry for continuous microparticle counting and sorting. Biosens. Bioelectron. **24**, 572–578 (2008)

93. Choi, W., et al.: Programmable manipulation of motile cells in optoelectronic tweezers using a grayscale image. Appl. Phys. Lett **93**(14), 143901 (2008)
94. Cheng, I.-F., et al.: Stepwise gray-scale light-induced electric field gradient for passive and continuous separation of microparticles. Microfluid. Nanofluid. **12**, 95–105 (2011)
95. Yang, S.-M., et al.: Dynamic manipulation and patterning of microparticles and cells by using TiOPc-based optoelectronic dielectrophoresis. Opt. Lett. **35**(12), 1959–1961 (2010)
96. Lu, Y.S., et al.: Controllability of non-contact cell manipulation by image dielectrophoresis. Opt. Quantum Electron. **37**, 1385–1395 (2005)
97. Choi, W., et al.: Lab-on-a-display: A new microparticle manipulation platform using a liquid crystal display (LCD). Microfluid. Nanofluid. **3**, 217–225 (2007)
98. Hwang, H., et al.: Interactive manipulation of blood cells using a lens-integrated liquid crystal display based optoelectronic tweezers system. Electrophoresis **29**, 1203–1212 (2008)
99. Hoeb, M., et al.: Light-induced dielectrophoretic manipulation of DNA. Biophys. J. **93**, 1032–1038 (2007)
100. Lee, M.-W., Lin, Y.-H., Lee, G.-B.: Manipulation and patterning of carbon nanotubes utilizing optically induced dielectrophoretic forces. Microfluid. Nanofluid. **8**, 609–617 (2010)
101. Iliescu, C., et al.: Fabrication of a dielectrophoretic chip with 3D silicon electrodes. J. Micromech. Microeng. **15**, 494–500 (2005)
102. Yu, L., et al.: Sequential field-flow cell separation method in a dielectrophoretic chip with 3D electrodes. J. Microelectromech. S. **16**, 1120–1129 (2007)
103. Iliescu, C., et al.: Bidirectional field-flow particle separation method in a dielectrophoretic chip with 3D electrodes. Sens. Actuators B **129**, 491–496 (2008)
104. Iliescu, C., Tresset, G., Xu, G.: Continuous field-flow separation of particle populations in a dielectrophoretic chip with three dimensional electrodes. Appl. Phys. Lett. **90**, 234104 (2007)
105. Iliescu, C., Tresset, G., Xu, G.: Dielectrophoretic field-flow method for separating particle populations in a chip with asymmetric electrodes. Biomicrofluidics **3**, 044104 (2009)
106. Tay, F.E.H., et al.: Electrical and thermal characterization of a dielectrophoretic chip with 3D electrodes for cells manipulation. Electrochim. Acta **52**, 2862–2868 (2007)
107. Park, B.Y., Madou, M.J.: 3D electrode designs for flow-through dielectrophoretic system. Electrophoresis **26**, 3745–3757 (2005)
108. Martinez-Duarte, R., Renaud, P., Madou, M.J.: A novel approach to dielectrophoresis using carbon electrodes. Electrophoresis **32**, 2385–2392 (2011)
109. Ma, W., et al.: High-throughput dielectrophoretic manipulation of bioparticles within fluids through biocompatible three-dimensional microelectrode array. Electrophoresis **32**, 494–505 (2011)
110. Wang, L., et al.: Dual frequency dielectrophoresis with interdigitated sidewall electrodes for microfluidic flow-through separation of beads and cells. Electrophoresis **30**, 782–791 (2009)
111. Wang, L., Flanagan, L., Lee, A.P.: Side-wall vertical electrodes for lateral field microfluidic applications. J. MEMS **16**(2), 454–461 (2007)
112. Wang, L., et al.: Dielectrophoresis switching with vertical sidewall electrodes for microfluidic flow cytometry. Lab Chip **7**, 1114–1120 (2007)
113. Zhang, Y.T., et al.: Titanium-based dielectrophoresis devices for microfluidic applications. Biomed. Microdevices **10**, 509–517 (2008)
114. Kang, Y., et al.: Continuous particle separation with localized AC-dielectrophoresis using embedded electrodes and an insulating hurdle. Electrochim. Acta **54**, 1715–1720 (2009)
115. Lewpiriyawong, N., et al.: Microfluidic characterization and continuous separation of cells and particles using conducting poly(dimethyl siloxane) electrode induced alternating current-dielectrophoresis. Anal. Chem. **83**, 9579–9585 (2011)
116. Xia, Y., Whitesides, G.M.: Soft lithography. Annu. Rev. Mater. Sci. **28**, 153–184 (1998)
117. Niu, X., et al.: Characterizing and patterning of PDMS-based conducting composites. Adv. Mater. **19**, 2682–2686 (2007)

118. Cetin, B., et al.: Continuous particle separation by size via AC-dielectrophoresis using a lab-on-a-chip device with 3-D electrodes. Electrophoresis **30**, 766–772 (2009)
119. Lewpiriyawong, N., Yang, C., Lam, Y.C.: Continuous sorting and separation of microparticles by size using AC dielectrophoresis in a PDMS microfluidic device with 3-D conducting PDMS composite electrodes. Electrophoresis **31**, 2622–2631 (2010)
120. Lewpiriyawong, N., Yang, C.: AC-dielectrophoretic characterization and separation of submicron and micron particles using sidewall AgPDMS electrodes. Biomicrofluidics **6**, 012807 (2012)
121. Lapizco-Encinas, B.H., et al.: Insulator-based dielectrophoresis for the selective concentration and separation of live bacteria in water. Electrophoresis **25**, 1695–1704 (2004)
122. Cummings, E.B., Singh, A.K.: Dielectrophoresis in microchips containing arrays of insulating posts: Theoretical and experiment results. Anal. Chem. **75**, 4724–4731 (2003)
123. Shafiee, H., et al.: Contactless dielectrophoresis: a new technique for cell manipulation. Biomed. Microdevices **11**, 997–1006 (2009)
124. Kang, Y., et al.: DC-Dielectrophoretic separation of biological cells by size. Biomed. Microdevices **10**, 243–249 (2008)
125. Lewpiriyawong, N., Yang, C., Lam, Y.C.: Dielectrophoretic manipulation of particles in a modified microfluidic H filter with multi-insulating blocks. Biomicrofluidics **2**, 034105 (2008)
126. Lewpiriyawong, N., Yang, C., Lam, Y.C.: Electrokinetically driven concentration of particles and cells by dielectrophoresis with DC-offset AC electric field. Microfluid. Nanofluid. **12**, 723–733 (2011)
127. Church, C., et al.: Electrokinetic focusing and filtration of cells in a serpentine microchannel. Biomicrofluidics **3**, 044109 (2009)
128. Chen, D., Du, H.: A microfluidic device for rapid concentration of particles in continuous flow by DC dielectrophoresis. Microfluid. Nanofluid. **9**, 281–291 (2010)
129. Chou, C.F., et al.: Electrodeless dielectrophoresis of single- and double-stranded DNA. Biophys. J. **83**, 2170–2179 (2002)
130. Lapizco-Encinas, B.H., Ozuna-Chacon, S., Marco, R.-P.: Protein manipulation with insulator-based dielectrophoresis and DC electric fields. J. Chromatogr. A **1206**, 45–51 (2008)
131. Jones, P.V., Staton, S.J.R., Hayes, M.A.: Blood cell capture in a sawtooth dielectrophoretic microchannel. Anal. Bioanal. Chem. **401**, 2103–2111 (2011)
132. Barrett, L.M., et al.: Dielectrophoretic manipulation of particles and cells using insulating ridges in faceted prism microchannels. Anal. Chem. **77**, 6798–6804 (2005)
133. Barbulovic-Nad, I., et al.: DC-dielectrophoretic separation of microparticles using an oil droplet obstacle. Lab Chip **6**, 274–279 (2006)
134. Kang, K.H., et al.: Continuous separation of microparticles by size with direct current-dielectrophoresis. Electrophoresis **27**, 694–702 (2006)
135. Zhu, J., et al.: DC dielectrophoretic focusing of particles in a serpentine microchannel. Microfluid. Nanofluid. **7**(6), 751–756 (2009)
136. Sabounchi, P., et al.: Sample concentration and impedance detection on a microfluidic polymer chip. Biomed. Microdevices **10**(5), 661–670 (2008)
137. Rosenthal, A., Voldman, J.: Dielectrophoretic traps for single-particle patterning. Biophys. J. **88**, 2193–2205 (2005)
138. Zhu, J., Xuan, X.: Dielectrophoretic focusing of particles in a microchannel constriction using DC-biased AC electric fields. Electrophoresis **30**, 2668–2675 (2009)
139. Hawkins, B.G., Kirby, B.J.: Electrothermal flow effects in insulating (electrodeless) dielectrophoresis systems. Electrophoresis **31**, 3622–3633 (2011)
140. Sridharan, S., et al.: Joule heating effects on electroosmotic flow in insulator-based dielectrophoresis. Electrophoresis **32**, 2274–2281 (2011)
141. Krishnan, R., et al.: Alternating current electrokinetic separation and detection of DNA nanoparticles in high-conductance solutions. Electrophoresis **29**, 1765–1774 (2008)

142. Demierre, N., et al.: Focusing and continuous separation of cells in a microfluidic device using lateral dielectrophoresis. Sens. Actuators B **132**, 388–396 (2008)
143. Shafiee, H., et al.: Selective isolation of live/dead cells using contactless dielectrophoresis (cDEP). Lab Chip **10**, 438–445 (2009)
144. Braschler, T., et al.: Continuous separation of cells by balanced dielectrophoretic forces at multiple frequencies. Lab Chip **8**, 280–286 (2008)
145. Tornay, R., et al.: Dielectrophoresis-based particle exchanger for the manipulation and surface functionalization of particles. Lab Chip **8**, 267–273 (2008)
146. Valero, A., et al.: A miniaturized continuous dielectrophoretic cell sorter and its applications. Biomicrofluidics **4**, 022807 (2010)
147. Nascimento, E.M., et al.: Dielectrophoretic sorting on a microfabricated flow cytometer: Label free separation of Babesia bovis infected erythrocytes. Bioelectrochemistry **73**, 123–128 (2008)
148. Piacentini, N., et al.: Separation of platelets from other blood cells in continuous-flow by dielectrophoresis field-flow-fractionation. Biomicrofluidics **5**, 034122 (2011)
149. Demierre, N., et al.: Characterization and optimization of liquid electrodes for lateral dielectrophoresis. Lab Chip **7**, 355–365 (2006)
150. Henslee, E.A., et al.: Selective concentration of human cancer cells using contactless dielectrophoresis. Electrophoresis **32**, 2523–2529 (2011)
151. Cui, H.-H., Lim, K.-M.: Pillar array microtraps with negative dielectrophoresis. Langmuir **25**, 3336–3339 (2009)
152. Jen, C.-P., Chen, W.-F.: An insulator-based dielectrophoretic microdevice for the simultaneous filtration and focusing of biological cells. Biomicrofluidics **5**, 044105 (2011)
153. Pethig, R.: Dielectrophoresis: Status of the theory, technology, and applications. Biomicrofluidics **4**, 022811 (2010)
154. Voldman, J.: Electrical forces for microscale cell manipulation. Annu. Rev. Biomed. Eng. **8**, 425–454 (2006)
155. Lee, D.-H., et al.: Dielectrophoretic particle–particle interaction under AC electrohydrodynamic flow conditions. Electrophoresis **32**, 2298–2306 (2011)
156. Yang, H., et al.: Low-energy electron-beam lithography of ZEP-520 positive resist. In: 1st IEEE International Conference on Nano/Micro Engineered and Molecular Systems (2006)
157. Cumming, D.R.S., et al.: Fabrication of 3 nm wires using 100 keV electron beam lithography and poly(methyl methacrylate) resist. Appl. Phys. Lett. **68**(3), 322 (1996)
158. Muller, T., et al.: A 3D microelectrode system for handling and caging single cells and particles. Biosens. Bioelectron. **14**, 247–256 (1999)
159. Green, N.G., Morgan, H.: Dielectrophoretic investigations of sub-micrometre latex spheres. J. Phys. D Appl. Phys. **30**, 2626–2633 (1997)
160. Holmes, D., et al.: On-chip high-speed sorting of micron-sized particles for high-throughput analysis. IEE Proc. Nanobiotech. **152**(4), 129–135 (2005)
161. Koo, O.K., et al.: Targeted capture of pathogenic bacteria using a mammalian cell receptor coupled with dielectrophoresis on a biochip. Anal. Chem. **81**, 3094–3101 (2009)
162. Goater, A.D., Burt, J.P.H., Pethig, R.: A combined travelling wave dielectrophoresis and electrorotation device: Applied to the concentration and viability determination of Cryptosporidium. J. Phys. D Appl. Phys. **30**, 65–69 (1997)
163. Bridlea, H., et al.: Detection of cryptosporidium in miniaturised fluidic devices. Water Res. **46**, 1641–1661 (2012)
164. Varshney, M., et al.: A label-free, microfluidics and interdigitated array microelectrode-based impedance biosensor in combination with nanoparticles immunoseparation for detection of Escherichia coli O157:H7 in food samples. Sens. Actuator B **128**, 99–107 (2007)
165. Kim, S., et al.: A microwire sensor for rapid detection of Escherichia coli K-12 in fresh produce. Innov. Food Sci. Emerg. **12**, 617–622 (2011)
166. Yagupsky, P., Nolte, F.: Quantitative aspects of septicemia. Clin. Microbiol. Rev. **3**, 269–279 (1990)

167. Sengupta, S., Gordon, J.E., Chang, H.-C.: Microfluidic diagnostic systems for the rapid detection and quantification of pathogens. In: Tien, W.-C., Finehout, E. (eds.) Microfluidics for Biological Applications, pp. 274–276. Springer, Berlin (2008)
168. Yang, L., Bashir, R.: Electrical/electrochemical impedance for rapid detection of foodborne pathogenic bacteria. Biotechnol. Adv. **26**, 135–150 (2008)
169. Guan, X., et al.: Rapid detection of pathogens using antibody-coated microbeads with bioluminescence in microfluidic chips. Biomed. Microdevices **12**, 683–691 (2010)
170. Yang, L.: Dielectrophoresis assisted immuno-capture and detection of foodborne pathogenic bacteria in biochips. Talanta **80**, 551–558 (2009)
171. Pratt, E.D., et al.: Rare cell capture in microfluidicdevices. Chem. Eng. Sci. **66**, 1508–1522 (2011)

Thermodynamic Analysis and Optimization Design of Heat Exchanger

Mingtian Xu, Jiangfeng Guo and Xuefang Li

Abstract In order to address the contradiction between the limited fossil fuel reserves and sharp increase of huge energy demand from the world economy and people's daily lives, there is an urgent need to develop energy saving measures. Heat exchanger as a device for heat transfer from one medium to another is widely applied in power engineering, petroleum refineries, chemical industries, food industries, and so on. Therefore it is of great value to improve the heat exchanger performance and save energy in heat exchange processes. Recently with the aim of reducing the unnecessary heat dissipation in heat exchange processes, we have studied thermodynamic analysis and optimization design of heat exchangers. Firstly based on the genetic algorithm and the improved entropy generation number which avoids the 'entropy generation paradoxes' induced by the original entropy generation number, we proposed an improved entropy generation minimization approach for heat exchanger optimization design. Secondly, we found that the entransy is a state variable and the second law of thermodynamics can be described by the entransy and entransy dissipation, this work places the entransy dissipation theory on a solid thermodynamic basis. Thirdly, based on the entransy dissipation theory we derived the expression of the local entransy dissipation rate for heat convection, developed variational principles for heat transfer and showed that this principle is compatible with the Navier–Stokes–Fourier equations. Fourthly, based on the entransy dissipation theory, we proposed a heat exchanger performance evaluation criterion called the entransy dissipation number and established a principle of entransy dissipation equipartition for heat exchanger optimization designs. Finally, we developed an entransy dissipation minimization approach for heat exchanger optimization design and applied it to the tube-and-shell heat exchanger optimization design.

M. Xu (✉) · J. Guo · X. Li
Department of Engineering Mechanics, School of Civil Engineering, Shandong University, Jinan 250061, People's Republic of China
e-mail: mingtian@sdu.edu.cn

L. Wang (ed.), *Advances in Transport Phenomena 2011*,
Advances in Transport Phenomena 3, DOI: 10.1007/978-3-319-01793-8_3,

1 Introduction

In this chapter the work presented in Refs. [1–11] is reviewed. The second law of thermodynamics in terms of entransy and entransy dissipation is applied in the closed and open thermodynamic systems.

2 An Improved Entropy Generation Minimization Approach for Plate-Fin Heat Exchanger Design

Usually the entropy generation number which was defined by dividing the entropy generation rate by the capacity flow rate is taken as the objective function for plate-fin heat changer optimization designs based on the second law of thermodynamics. However, it is well-known that the entropy generation number causes some paradoxes and inconsistencies. Recently, a revised entropy generation number was proposed by Hesselgreaves, which is a dimensionless entropy generation non-dimensionalised by the ratio of heat flow to the input temperature of cold fluid. It was shown that the revised entropy generation number avoids the "entropy generation paradoxes". In this section based on the genetic algorithm an improved entropy generation minimization approach with the aim of minimizing the revised entropy generation number is developed for plate-fin heat exchanger optimization designs with multiple design variables. We find that in comparison with the randomly generated initial design the improved entropy generation minimization approach not only decrease the total fan power, but also improve the exchanger effectiveness. Finally this improved entropy generation minimization approach is extended to the optimization design of a waste heat recovery ventilation system where the heat exchanger works as a component.

2.1 Introduction

Last decades witness the rapid development of thermodynamic analysis and optimization design of heat exchangers in the framework of the second law of thermodynamics [12–17]. The highlight is the entropy generation minimization (EGM) approach developed by Bejan [12–16]. The essence of this approach is to minimize the entropy generation that describes the thermodynamic irreversibilities induced by heat conduction across the finite temperature difference, fluid friction and fluid mixing, which are detrimental to the heat exchanger performance. Based on the concept of entropy generation, several heat exchanger performance criteria were proposed, such as the entropy generation rate [13], entropy generation number (EGN) defined by dividing the entropy generation rate by the capacity flow rate $\dot{m}c_p$ [12, 18], augmentation entropy generation number [12], heat exchanger reversibility norm [19], Witte and Shamsundar efficiency [20] and local entropy generation number [21]. These criteria have their own features, but also are interconnected, as discussed in [17]. Among them the most widely applied is the EGN developed by Bejan [12, 14, 18]. With the EGN as the objective function, Bejan proposed an optimization method for counter-flow heat exchanger designs and used it to obtain the minimum heat transfer area of a shell and tube regenerative heat exchanger when the amount of heat transfer units was fixed [16, 22]. Grazzini and Gori [23], Sekulic [19], Zhang et al. [24], Ordonez and Bejan [25], and Bejan [26, 27] showed that subject to some constraint conditions the geometries of heat exchangers can be optimized through minimizing the irreversibilities occurring in heat exchangers. In Ref. [28] an analytical method for the EGM was developed to optimize the dimensions of various fin configurations. Bejan [14] showed that the EGM may play an important role in the preliminary stage of design in order to identify trends and the existence of optimization opportunities. Vargas et al. [29] presented an approach to determining the internal geometric configuration of a tube bank by optimizing the global performance of the environmental control system of a modern aircraft where the cross-flow heat exchanger works as a component. In Ref. [30] a balanced cross-flow plate type heat exchanger operating with unmixed fluids was analyzed based on the second law of thermodynamics, the variations of the minimum entropy generation number with respect to the flow path length, dimensionless mass velocity, dimensionless heat transfer area and dimensionless heat transfer volume were investigated. Reddy et al. [31] derived an expression of the EGM for a waste heat recovery steam generator which consists of an economizer, an evaporator and a superheater, and studied the influences of various non-dimensional operating parameters on the EGN. In Ref. [32] the influence of the heat exchanger flow length and the pin length on the EGN was explored, it was found that a larger number of passages with smaller pin height in the given frontal area of heat exchanger are more advantageous than less heat exchanger passages with larger pin height. In Ref. [33] an optimization method based on the genetic algorithm was developed for the cross-flow plate-fin heat exchanger design with the aim at minimizing the EGN.

Although the EGN has been widely applied in the optimization of heat exchanger designs to quantify the irreversibilities occurring in the heat exchanger, it suffers from the ‘entropy generation paradoxes’ [34]. In order to resolve this paradox, an alternate dimensionless entropy generation number was defined by dividing the entropy generation rate by Q/T_e (Q is the heat transfer rate and T_e is the ambient temperature) in Refs. [20, 35]. However, a new parameter T_e introduced in this approach causes further complication [17]. Recently a more appropriate means of non-dimensionalising the entropy generation by $Q/T_{c,i}$ ($T_{c,i}$ is the inlet temperature of the cold fluid) was proposed by Hesselgreaves [34], the dimensionless entropy generation number defined in such a way was called as the revised entropy generation number (REGN) and denoted as N_{s1} in the following discussion. In this subsection, with the REGN taken as the objective function, a multivariable optimization problem for the plate-fin heat exchanger design is first formulated and then solved by the genetic algorithm. Finally this approach is extended to optimize the design of a waste heat recovery ventilation system in which the heat exchanger works as a component.

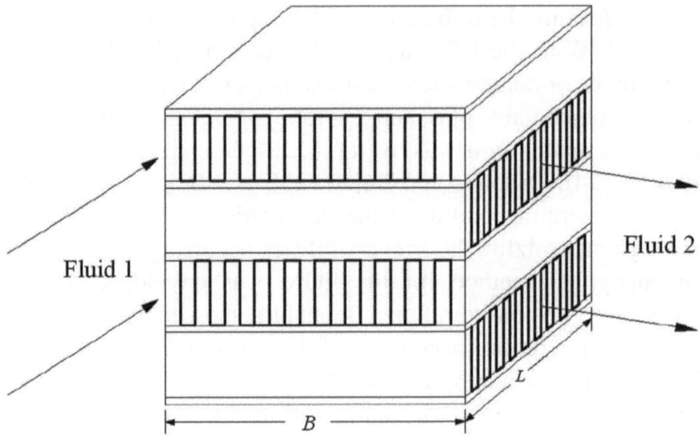

Fig. 1 Plate-fin heat exchanger

2.2 The Improved EGM Approach for Plate-Fin Heat Exchangers

The plate-fin heat exchanger depicted in Fig. 1 finds wide applications in various industrial fields [36]. In this subsection, with the REGN as the objective function an improved entropy generation minimization approach is developed and applied to the optimization of the cross-flow plate-fin heat exchanger design. In the following discussion the conventional assumptions for the heat exchanger design are adopted, such as no longitudinal heat conduction, negligible potential and kinetic energy changes, negligible heat transfer between the exchanger and its surroundings and so on [37].

2.2.1 Heat Transfer and Pressure Drop in Plate-Fin Heat Exchanger

For a cross-flow plate-fin heat exchanger the Reynolds number is usually defined as

$$Re = \frac{G \cdot d_e}{\mu}, \tag{1}$$

where G is the fluid mass velocity, d_e is the hydraulic diameter of flow passages, and μ is dynamic viscosity. The heat transfer correlation in terms of heat transfer factor j for plate-fin heat exchangers reads [36]

$$j = St \cdot Pr^{2/3} \tag{2}$$

where St is Stanton number, and Pr is Prandtl number. Kays and London have documented plots of the heat transfer factor j and fanning friction factor f versus Reynolds number for the tube tank, tube-fin heat exchangers, and other plate-fin surface geometries in Ref. [38]. And many heat transfer correlations for plate-fin

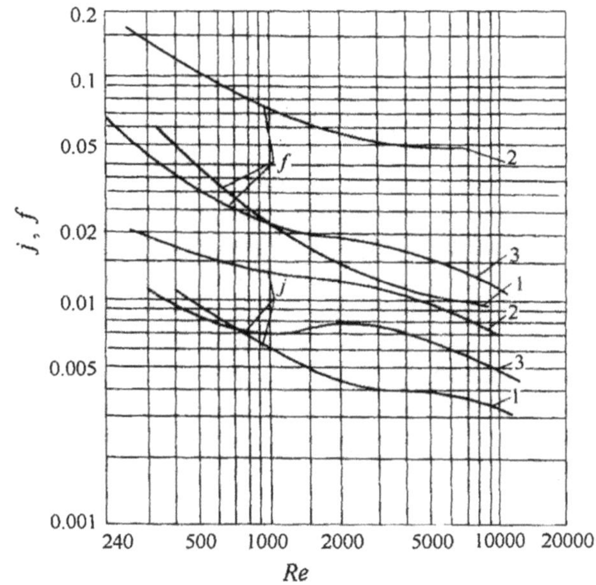

Fig. 2 Performance *curves* of fins with three kinds of geometries introduced by KOBELCO in Japan [39]. *1* rectangular fin, *2* offset strip fin, *3* perforated fin

heat exchangers are available in Ref. [39]. Among them the fin performance curves introduced by KOBELCO in Japan are widely applied in China, and the reliability of the data has been examined by practice. These curves are shown in Fig. 2. When $300 \leq Re \leq 7,500$ the corresponding correlations for the plate-fin heat exchangers with offset strip fins are written as [39]

$$\ln j = -2.64136 \times 10^{-2}(\ln Re)^3 + 0.555843(\ln Re)^2 - 4.09241 \ \ln Re + 6.21681 \tag{3}$$

$$\ln f = 0.132856(\ln Re)^2 - 2.28042(\ln Re) + 6.79634. \tag{4}$$

The heat transfer coefficient α in one side of the fluids is written as follows

$$\alpha = St \cdot c_p \cdot G = f(Re)\frac{c_p \cdot G}{Pr^{2/3}} \tag{5}$$

where c_p is the specific heat at constant pressure, G is the mass flow velocity. When the heat capacity rate ratio C^* is equal to one, the entropy generation rate for the cross-flow heat exchanger becomes [30]

$$\dot{S}_{\text{gen}} = \dot{m}c_p \ \ln\frac{T_{\text{h,o}}}{T_{\text{h,i}}} + \dot{m}c_p \ \ln\frac{T_{\text{c,o}}}{T_{\text{c,i}}} - \dot{m}R \ \ln\left(1 - \frac{\Delta P_{\text{h}}}{P_{\text{h,i}}}\right) - \dot{m}R \ \ln\left(1 - \frac{\Delta P_{\text{c}}}{P_{\text{c,i}}}\right) \tag{6}$$

where $\dot{m}$ is the mass flow rate, T is the temperature, R is the ideal gas constant, P is the pressure, ΔP is the pressure drop, the subscripts h and c respectively refer to the hot and cold fluids, the subscripts i and o represent the inlet and outlet of heat exchanger, respectively. The REGN N_{s1} is defined by dividing $\dot{S}_{gen}$ by $Q/T_{\text{c,i}}$,

$$N_{s1} = \frac{1}{\varepsilon(1/\tau - 1)}\{\ln[1 - \varepsilon(1/\tau - 1)] + \ln[1 + \varepsilon(1/\tau - 1)]\} \\ - \frac{R}{\varepsilon c_p(1/\tau - 1)}\ln\left(1 - \frac{\Delta P_h}{P_{h,i}}\right) - \frac{R}{\varepsilon c_p(1/\tau - 1)}\ln\left(1 - \frac{\Delta P_c}{P_{c,i}}\right) \tag{7}$$

where $\tau = T_{c,i}/T_{h,i}$ is the ratio of inlet temperature of the cold fluid to inlet temperature of the hot fluid, ε is the heat exchanger effectiveness. The first line of the right side of Eq. (7) represents the dimensionless entropy generation N_{s1T} related to the heat conduction across finite temperature difference; last two terms stand for the dimensionless entropy generation N_{s1F} related to the flow friction. When $C^* = 1$ the exchanger effectiveness for the single-pass unmixed–unmixed cross-flow heat exchanger is expressed as [36],

$$\varepsilon = 1 - \exp\{C^* Ntu^{0.22}[\exp(-C^* Ntu^{0.78}) - 1]\} \tag{8}$$

where Ntu is the number of exchanger heat transfer units, C^* is the heat capacity flow rate ratio.

For the pressure drop in the plate-fin exchanger, we neglect the entrance, momentum and exit effects. Therefore, the pressure drop and the number of heat transfer units are written as [12, 30],

$$\frac{\Delta P}{P} = f\frac{4L}{d_e}\frac{G^2}{2\rho P} \tag{9}$$

$$Ntu = \frac{4L}{d_e}St = j\frac{4L}{d_e}Pr^{-2/3} \tag{10}$$

where ρ is the fluid density, L is the length of flow passage. Substituting Eqs. (8), (9) and (10) into Eq. (7) gives rise to the final expression of the REGN which describes the dependence of the REGN on design parameters.

Now let us consider a specific gas-to-gas plate-fin heat exchanger. The known parameters are documented in Table 1. The variation of the REGN with respect to the ratio of the flow passage length to the hydraulic diameter (L/d_e) is shown in Fig. 3. From this figure one can see that the dependences of N_{s1T} and N_{s1F} on L/d_e demonstrate the opposite behaviors, therefore there exists an optimal value of L/d_e which minimizes the REGN. This indicates that the appropriate selection of the values of the parameters of the heat exchanger can lead to the minimum entropy generation.

2.2.2 Optimization of Plate-Fin Heat Exchanger Design

In our heat exchanger optimization design approach in the framework of the second law of thermodynamics, the REGN defined by Eq. (7) is taken as the objective function, the admissible pressure drop and the requirements from the design standard are set to the constraint conditions and some geometrical

Table 1 Known data for a gas-to-gas heat exchanger

Parameters	Values
Inlet temperature of hot fluid $T_{h,i}$ (°C)	80
Inlet temperature of cold fluid $T_{c,i}$ (°C)	20
Inlet pressure of two streams P_i (MPa)	0.45
Specific heat of two fluids c_p (J/(kg K))	927.5
Dynamic viscosity of two fluids μ (Pa s)	23.36×10^{-6}
Density for two fluids ρ (kg/m^3)	5.4493
Ideal gas constant for air R (J/(kg K))	287
Mass flow velocity G (kg/(m^2 s))	20

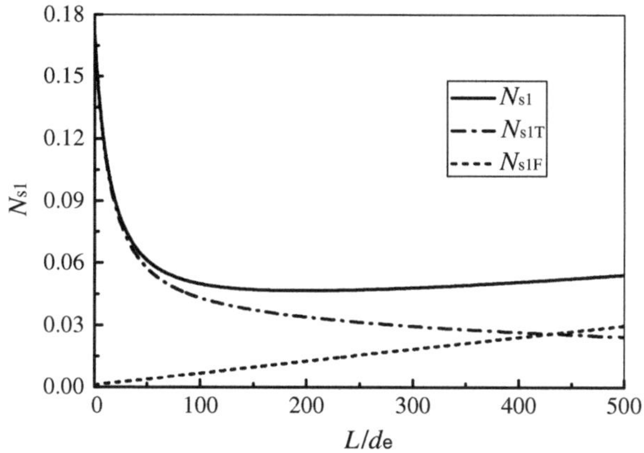

Fig. 3 Variations of the revised EGN with respect to L/d_e ($Re = 3{,}500$)

parameters are selected as the design variables. In the following discussions, for a specific plate-fin heat exchanger optimization design problem, more details about the optimization requirements and the design variables will be given. After the optimization design problem of the plate-fin heat exchanger is formulated, we should select an appropriate numerical method to solve the optimization problem. One may use the traditional approaches which require the information of the gradients of objective functions. Unfortunately they suffer from getting trapped at local optimums and can't ensure that the global optimal solution is solved [40]. In comparison with other approaches, the genetic algorithm (GA) demonstrates appealing features in solving optimization problems. First it provides a high level of robustness by simulating nature's adaptation in the evolution process [41]. More importantly, the GA has very powerful capability to find the global optimum [42]. Therefore, the GA is widely applied to seek the global optimal solution of the optimization problem [43]. In the following, the GA is employed to solve the plate-fin heat exchanger optimization design problems.

The GA is based on the natural selection rules which were found in biological evolution processes. In tackling the optimization problem, before a GA is put to work, the potential solutions should be encoded in a form that a computer can process. A potential solution is called an individual and can be represented by a set of parameters, which can be encoded values in binary form as the genes of a chromosome. Initial population of individuals is formed from a random set of solutions, and then next generations are generated through some operators discussed in the following.

A metric called a *fitness function* that allows each potential solution (individual) to be quantitatively evaluated. After a random initial population in the ranges of design variables is generated, the algorithm creates a sequence of new generations iteratively until the stopping criterion is met. In this process, offspring are generated by merging two individuals in the current generation through a *crossover operator*, or by modifying a chromosome with a *mutation operator*. A new generation is formed by some parents and offspring based on fitness values, the population size is kept constant by eliminating the inferior ones. The chromosomes with higher fitness values have higher probabilities to survive; this guarantees the algorithm converges to a best individual after certain generations, which probably represents the best solution of the given problem [44, 45]. The flow chart of a genetic algorithm is shown in Fig. 4.

In order to validate the reliability and accuracy of the GA, two typical test examples are considered. The first one is to search the global minimum point of the following function,

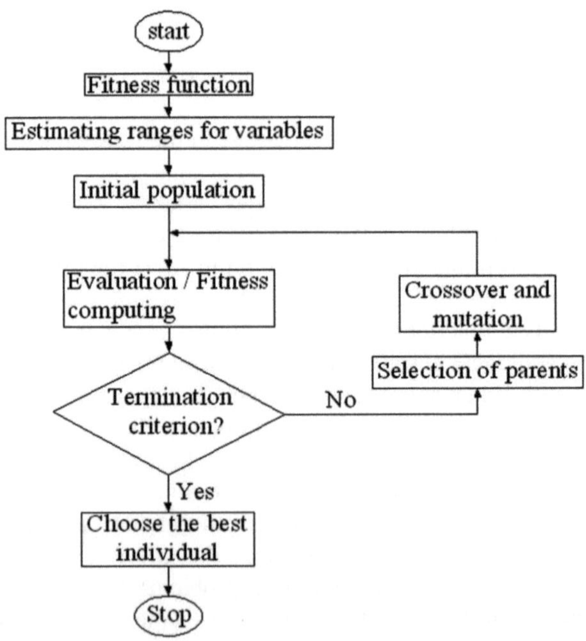

Fig. 4 Flow chart of genetic algorithm

$$f(x) = \sum_{i=1}^{5} x_i^2, \quad -512 \leq x_i \leq 512 \tag{11}$$

where $x_i(i = 1, 2, \cdots, 5)$ are variables. The function $f(x)$ is a continuous unimodal function; it has only one minimum point located at (0, 0, 0, 0, 0), and the minimum value is equal to 0. The initial population is generated randomly within the ranges of parameters, whose size is set to 20 and remains the same in the searching process. The termination condition is that the number of generations is not greater than 500. The global optimal solution obtained by this approach is shown in Table 2. Evidently, the GA has very high accuracy.

For the second example, we consider the Shaffer's F_6 function defined as follows:

$$f(x,y) = 0.5 - \frac{\sin^2\left(\sqrt{x^2+y^2}\right) - 0.5}{\left(1+0.001(x^2+y^2)\right)^2}, \quad -100 < x, y < 100 \tag{12}$$

This function has infinite local maximum values, but it has only one global maximum value located at (0, 0) as shown in Fig. 5. The global maximum value is surrounded by a circle of local maximum values, which are all equal to 0.9903. Hence when searching the global maximum point, it is quite easy to be trapped at the local maximum points. The GA is adopted to solve this function, the size of population is set to 20, and the maximum number of generations is set to 500. The searching process and the final result are documented in Table 3. One can see that

Table 2 The global optimal solution of the function $f(x)$

x_1	x_2	x_3	x_4	x_5	$f(x)$
-6.05×10^{-7}	-1.14×10^{-8}	2.2×10^{-8}	-8.60×10^{-7}	3.16×10^{-7}	1.21×10^{-12}

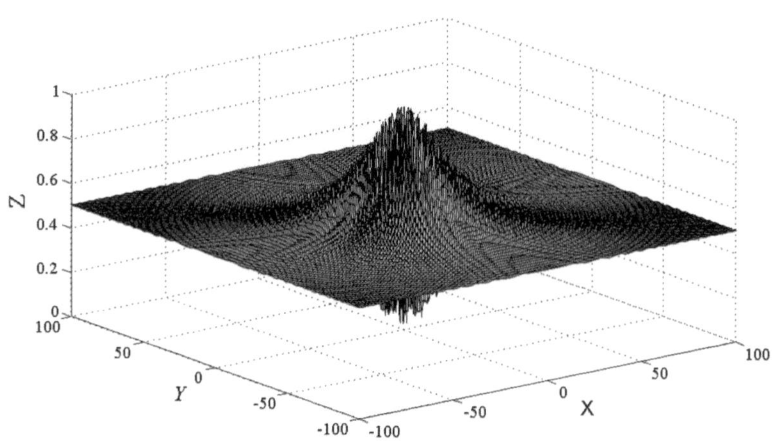

Fig. 5 The three-dimensional image of Shaffer's F_6 function

Table 3 The solving process of the GA for the Shaffer's F_6 function

Generations	x	y	$f(x, y)$
1	33.2214	10.2575	0.5925
50	7.2035	6.2864	0.9043
183	−0.1170	−3.0949	0.9886
184	−0.1058	−3.1455	0.9902
196	−0.0180	−3.1334	0.9903
325	0.0010	0.0476	0.9977
340	0.0008	−0.0022	1.0000
500	0.0000	0.0000	1.0000

the GA has not got trapped at local optimum points and eventually gives rise to the exact global optimal solution. This example shows that the GA is a reliable and powerful tool to tackling the global optimization problems and usually has high accuracy.

Example One

In this example we consider the optimization design of a cross-flow plate-fin type heat exchanger with offset strip fin. Assume that the hot and cold fluids are balanced, namely $C^* = 1$. The known data for the plate-fin heat exchanger under consideration has been documented in Table 1. The ranges of the design variables and the constraint conditions are given in Table 4 according to the national standards of China [39]. Based on the data in Tables 1 and 4, the other parameters of the heat exchanger can be determined. The hydraulic diameter is given as,

$$d_e = \frac{2xy}{x + y} \tag{13}$$

where x is the internal span of fins, the inside height of fins $y = H - \delta$, δ is the thickness of fin and is set to 0.3 mm in the following calculation. The effective cross section area in one layer f_i is expressed as,

Table 4 Design variables as well as their ranges and constraint conditions for example one

Variables	Ranges
Design variables	
The fin height H (mm)	3.0–12.0
The fin pitch x (mm)	1.4–4.2
The number of hot side layers n	20–80
The length of heat exchanger core L (m)	0.01–2.0
Constraint conditions	
Pressure drop $\triangle P_h$ in hot side (Pa)	<5,000
Pressure drop $\triangle P_c$ in cold side (Pa)	<6,000
Re in two sides	300 < Re < 7,500

$$f_{\mathrm{i}} = \frac{L}{s} xy \tag{14}$$

In this example, both fluids are assumed as ideal gases and they have the same mass flow rates. So the effective width is equivalent to the length of flow channel. Then the heat transfer area of one layer F_i can be written as follows,

$$F_{\mathrm{i}} = 2(x+y)L^2/s. \tag{15}$$

The mass flow velocity in the channel reads

$$G = \frac{\dot{m}}{nf_{\mathrm{i}}} \tag{16}$$

Since the thickness of fin is usually far less than the fin offset length, the transverse heat conduction across the fin is usually neglected, and the staggered arrangement for hot and cold streams is assumed, then the fin efficiency η_f is expressed as [38],

$$\eta_f = \frac{\tan\left(\sqrt{\frac{2\alpha}{\lambda_{\mathrm{w}}\delta}}\frac{H}{2}\right)}{\sqrt{\frac{2\alpha}{\lambda_{\mathrm{w}}\delta}}\frac{H}{2}} \tag{17}$$

where λ_{w} is the thermal conductivity of fins and equals to 190 W/(m K), and H is the height of fin. So the total fin effective efficiency η_o is

$$\eta_{\mathrm{o}} = 1 - \frac{y}{x+y}(1-\eta_f) \tag{18}$$

If the wall and fouling thermal resistances are negligible, the total heat transfer coefficient K_h based on the hot fluid side is

$$K_{\mathrm{h}} = \left[\frac{1}{\alpha_{\mathrm{h}}\eta_{\mathrm{oh}}} + \frac{1}{\alpha_{\mathrm{c}}\eta_{\mathrm{oc}}}\frac{F_{\mathrm{h}}}{F_{\mathrm{c}}}\right]^{-1} \tag{19}$$

where F_{h} and F_{c} denote the heat transfer areas in the hot and cold fluid sides, respectively. The number of heat transfer units based on the hot fluid side is expressed as,

$$Ntu = \frac{K_{\mathrm{h}}F_{\mathrm{i}}n}{\dot{m}c_p} \tag{20}$$

The heat transfer rate is written as follows,

$$Q = \dot{m}c_p\varepsilon\left(T_{\mathrm{h,i}} - T_{\mathrm{c,i}}\right) \tag{21}$$

where the exchanger effectiveness ε is expressed by Eq. (8). The total power of fans is [46]

$$W = \frac{1}{\eta}\left(\frac{\dot{m}_{\mathrm{h}}}{\rho_{\mathrm{h}}}\Delta P_{\mathrm{h}} + \frac{\dot{m}_{\mathrm{c}}}{\rho_{\mathrm{c}}}\Delta P_{\mathrm{c}}\right) \tag{22}$$

where η is the fan efficiency. Furthermore, the following *JF* factor is defined for evaluating the heat exchanger performance [47]:

$$JF = \frac{j/j_R}{(f/f_R)^{1/3}} \tag{23}$$

where j_{R} and f_{R} are the reference values of j and f factors, respectively.

Based on the known data of the heat exchanger under consideration and the above heat transfer calculation, we attempt to optimize the plate-fin heat exchanger design. The objective function is given by Eq. (7). The numbers of the initial population and maximum generation are set to 20 and 500, respectively. The whole process of searching the global optimum solution of the optimization design problem by the GM is described as follows:

Step 1: Choose the fitness function. In the following calculation the minus REGN is chosen as the fitness function.

Step 2: Determine the design variables and the ranges of their values as well as the constraint conditions. The selected design variables and the ranges of their values as well as the constraint conditions are listed in Table 4.

Step 3: Once the values of design variables are given, the other parameters of the heat exchanger can be calculated by the traditional heat exchanger design process as discussed above.

Step 4: Calculate the value of fitness function for every individual (potential solution), and judge whether it violates the constraint conditions, if so the fitness of this individual is set to a very small value. After that, check if the termination condition is met. If so the individual with the maximum value of the fitness function is set to the global optimum solution of the optimization design problem of the plate-fin heat exchanger, otherwise, move to the next step.

Step 5: Choose the parents among all the individuals based on their fitness function values calculated in step 4 and generate next generation through *cross operator* and *mutation operator*. Then return to step 4.

For the details about the GA, please refer to Refs. [43–45]. Based on the genetic algorithm toolbox in Matlab [43] we develop a code for solving the optimization design problem of the plate-fin heat exchanger. After 500 generations of evolution, the process is terminated and the best individual (optimum solution) is obtained. For illustrating the evolution process, the variation of the REGN of the best individual in every generation with respect to the number of generations is displayed in Fig. 6. This figure shows that with increasing the number of generations the REGN declines. When the number of generations exceeds 100, the REGN tends to be a constant. Hence the iterative process is convergent. The parameters corresponding to the initial design plan and the optimum result are documented in Table 5. From this table one can see that the optimal design solution reduces the power of fans by 63.3 % and increases the exchanger effectiveness by 4.13 % in comparison with a randomly generated initial design of the heat exchanger. This is achieved by appropriately selecting the fin height, the number of layers, the flow passage length and the fin pitch. The variations of the exchanger effectiveness, the pressure drop, the number of the heat transfer units and *JF* factor of the best individual for every generation during the evolution process are displayed in Fig. 7. With decreasing the REGN Fig. 7a shows that the overall trend is to enhance the exchanger effectiveness, Fig. 7b indicates that the power of fans is significantly decreased. Figure 7c tells us that when the iterative process proceeds, the overall tendency of the number of heat transfer units is to increase. From Fig. 7d one can see that the *JF* factor is larger than one, and become larger and larger with decreasing N_{s1}, thus the heat exchanger performance becomes better and better [47], and reaches the best when N_{s1} takes its minimum value. Note that although the initial design is selected randomly, they satisfy the heat load requirement, the constraint conditions and the design standards, therefore they are the eligible heat exchanger design plans. Furthermore it exhibits similar behavior for different initial design to approach the optimum design through the optimization process.

Example Two

In this example we consider the optimization design of a cross-flow plate-fin heat exchanger with given heat load. In this heat exchanger the hot and cold fluids are oxygen and air, respectively. The known information of the heat exchanger under consideration is documented in Table 6 (the symbol '–' represents the undetermined data). The design variables and their ranges of the allowable values are listed in Table 7. Note that the heat capacity flow rates of the hot and cold fluids are not same.

For the cross-flow heat exchanger where both fluids are unmixed, although the series expression of the effectiveness is available [36], it is not convenient to use.

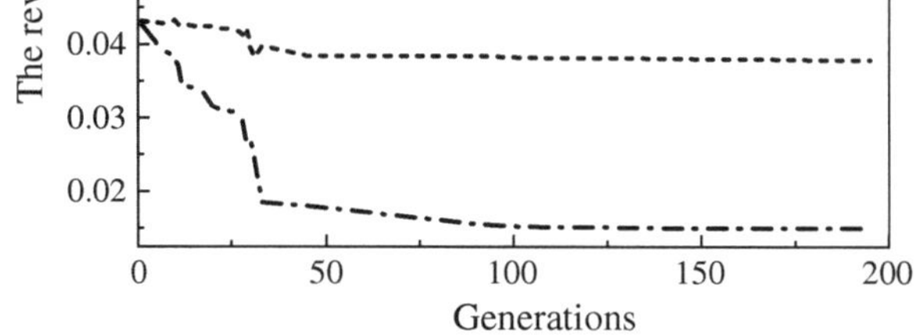

Fig. 6 The REGN versus the number of generations

Table 5 The initial design plan and optimum result for the plate-fin heat exchanger

	H (mm)	x (mm)	n	L (m)	d_e (mm)	Re	Ntu	ε	W (kW)	N_{s1H}	N_{s1F}	N_{s1}
Initial	9.5	4.1	72	1.582	5.7	7,458.6	5.41	0.7454	56.6	0.0431	0.0426	0.0857
Finial	12.0	4.1	80	1.998	6.1	4,482.6	7.78	0.7762	20.78	0.0379	0.015	0.0529

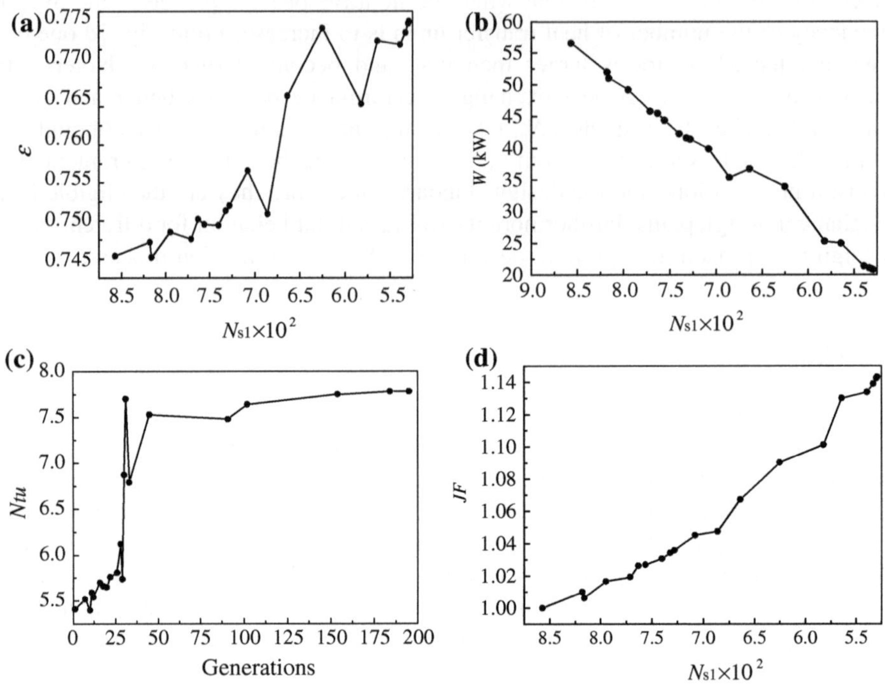

Fig. 7 Variations of the exchanger effectiveness, power of fans, number of heat transfer units and JF factor during the iterative process of genetic algorithm. **a** The exchanger effectiveness versus N_{s1}, **b** the power of fans versus N_{s1}. **c** Variation of Ntu with the number of generations, **d** JF factor versus N_{s1}

Table 6 The known data of the heat exchanger with the fixed heat load for example two

Parameters	Hot gas (oxygen)	Cold gas (air)
Inlet temperature T_i (°C)	80	20
Outlet temperature T_o (°C)	30	–
Mass flow rate $\dot{m}$ (kg/s)	20	30
Density ρ (kg/m^3)	5.4493	1.2490
Specific heat at constant pressure c_p (J/(kg K))	927.5	998.5
Dynamic viscosity μ (Pa s)	23.36×10^{-6}	18.225×10^{-6}
Inlet pressure P (MPa)	0.45	0.11
Thermal conductivity λ (W/(m K))	0.03015	0.02582

Table 7 Design variables as well as their value ranges and constraint conditions for example two

Variables	Ranges
Design variables	
The fin height in hot side H_h (mm)	3.0–12.0
The fin height in cold side H_c (mm)	3.0–12.0
The fin pitch in hot side x_h (mm)	1.4–4.2
The fin pitch in cold side x_c (mm)	1.4–4.2
The number of hot side layers n	20–100
The effective width B (m)	0.01–1.5
Constraint conditions	
Pressure drop $\triangle P_h$ in hot side (Pa)	<5,000
Pressure drop $\triangle P_c$ in cold side (*Pa*)	<6,000
Re in two sides	300 < Re < 7,500

Therefore, in our treatment, the following approximate formula is developed by curve fitting based on the data presented in [39],

$$\varepsilon = \frac{1.1054 - 1.0657Ntu + 0.1215C^{*} - 0.2981C^{*2}}{1 - 0.9483Ntu + 0.2880C^{*} - 0.3683C^{*2}} \tag{24}$$

The heat load is expressed as

$$Q = \left(\dot{m}c_p\right)_{\rm h}\left(T_{\rm h,i} - T_{\rm h,o}\right) \tag{25}$$

The flow passage length is written as follows

$$L = \frac{Q}{K_{\rm h}\Delta t_{\rm lm}2n(x_{\rm h} + y_{\rm h})B/s_{\rm h}} \tag{26}$$

where $\Delta t_{\rm lm}$ is the log-mean temperature difference. For the convenience of comparison, all parametersare calculated based on the hot fluid side. Then the similar optimization process as described in Example One is implemented. The variations of the REGN, *Ntu* and fan power *W* with respect to the number of generations of the evolution process are shown in Fig. 8. In this example the heat load and the mass flow rate are fixed as shown in Table 6. Hence the exchanger effectiveness is fixed, which can also be seen from the $N_{\rm s1T}$-curve in Fig. 8a and *Ntu*-curve in Fig. 8b. However, the entropy generation related to the fluid friction experiences a sharp decrease at the early stage of the iterative process as shown in Fig. 8a. The similar behavior occurs for the total fan power as shown in Fig. 8b. The total fan power is decreased by 78 % through the optimization process. The *JF* factor defined by Eq. (23) increases from 1 to 1.5. The comparison between an initial design plan and the optimal solution are shown in Table 8. From this table, one can see that the exchanger effectiveness and the revised entropy generation number for heat conduction N_{s1H} respectively takes the constant values 0.6 and 0.0556, while the power of fans and the revised entropy generation number for fluid friction N_{s1F} fall to 36 kW and 0.0273 from 165 kW and 0.1271, respectively.

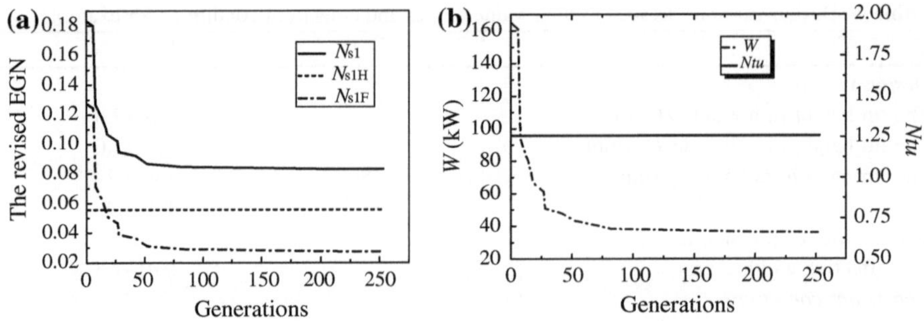

Fig. 8 Variations of the REGN, *Ntu* and *W* with respect to the number of generations. **a** The revised EGN versus the number of generations. **b** *Ntu* and *W* versus the number of generations

Table 8 The initial design plan and optimum solution for example two

	H_h (mm)	H_c (mm)	x_h (mm)	x_c (mm)	n	B (mm)	Ntu	ε	W (kW)	N_{s1H}	N_{s1F}	N_{s1}
Initial	12.0	11.0	2.4	1.8	95	1.34	1.26	0.6	165	0.0556	0.1271	0.1827
Finial	12.0	12.0	1.9	4.2	100	1.5	1.26	0.6	36	0.0556	0.0273	0.0829

Therefore the overall performance of the heat exchanger is significantly improved by the optimization design.

Example Three

In this subsection the application of the improved entropy generation minimization approach at the system level is investigated. More specifically, we consider the optimization design of a waste heat recovery ventilation system in which the heat exchanger works as a component. Assume that the room temperature is maintained at 20 °C and the outdoor temperature is −5 °C. Therefore the outdoor air needs to be heated from −5 to 20 °C before going into the room. The known data of the system is documented in Table 9. If the cold air is heated by electric heater, the annual power consumption C_H can be calculated as follows:

$$C_{\mathrm{H}} = \left(\dot{m}c_p\right)_{\mathrm{c}}\left(T_{\mathrm{r}} - T_{\mathrm{c,o}}\right)t \tag{27}$$

where T_r is the room temperature, t is the operating hours in one year and is set to 4,000 h/year in our calculation. Thus the annual power consumption C_{H} is 10,050 kWh. In order to recycle the heat of the exhausted air from the room, before being discharged from the room to the environment we let it first go into a counter-flow plate-fin heat exchanger where the exhausted hot air and the cold air can exchange heat. Such a waste heat recovery ventilation system is depicted in Fig. 9. The information of the system is documented in Table 9. Based on this known system information the annual electrical energy consumptions for heating

Table 9 The known data for the waste heat recovery ventilation system

Parameters	Hot air	Cold air
Inlet temperature T_i (°C)	20	−5
Mass flow rate $\dot{m}$ (kg/s)	0.1	0.1
Density ρ (kg/m^3)	1.247	1.270
Specific heat at constant pressure c_p (J/(kg K))	1,005	1,005
Dynamic viscosity μ (Pa s)	17.6×10^{-6}	17.4×10^{-6}
Inlet pressure P (Pa)	101,325	101,325
Thermal conductivity λ (W/(m K))	0.0251	0.02475

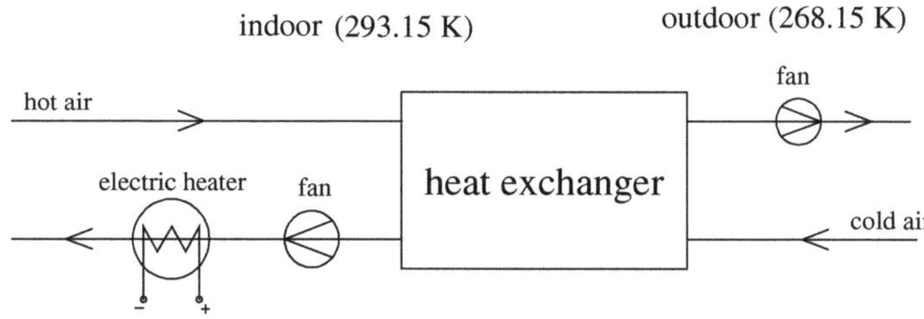

Fig. 9 Waste heat recovery ventilation system

the cold air by electric heater and for driving fans are equal to 2,492 and 13.48 kWh, respectively. Therefore the reduction of the annual power consumption of the system by using the counter-flow plate-fin heat exchanger to recover the waste heat is 7,544 kWh.

In order to save more energy, the system needs to be optimized. Assume that the heat exchange area is fixed at 20 m^2, and the power of fans cannot exceed 30 W, the entropy generation related to the fluid mixing is neglected, and the heat transfer between the heat exchanger and environment is negligible. Then the total entropy generation number of the waste heat recovery ventilation system is expressed as:

$$\begin{aligned} \dot{S}_{\text{gen}} &= \dot{S}_{\text{g,H}} + \dot{S}_{\text{g,T}} + \dot{S}_{\text{g,F}} + \dot{S}_{\text{g,E}} \\ &= \left(\dot{m}c_p\right)_{\text{c}} \ln\frac{T_{\text{r}}}{T_{\text{c,o}}} + \left(\dot{m}c_p\right)_{\text{h}} \ln\frac{T_{\text{h,o}}}{T_{\text{h,i}}} + \left(\dot{m}c_p\right)_{\text{c}} \ln\frac{T_{\text{c,o}}}{T_{\text{c,i}}} - \dot{m}_{\text{h}} R \ln\left(1 - \frac{\Delta P_{\text{h}}}{P_{\text{h,i}}}\right) \\ &\quad - \dot{m}_{\text{c}} R \ln\left(1 - \frac{\Delta P_{\text{c}}}{P_{\text{c,i}}}\right) + \left(\dot{m}c_p\right)_{\text{h}} \frac{\left(T_{\text{h,o}} - T_{\text{e}}\right)}{T_{\text{e}}} \end{aligned} \tag{28}$$

where T_r is the room temperature. The first term in Eq. (28) is the rate of entropy generation due to cold air heating by the electric heater, the last term accounts for the rate of entropy generation due to heat transfer from the hot air leaving the heat exchanger to the environment at the temperature T_{e}, and the other terms account for the rate of entropy generation related to the heat conduction across finite temperature difference and fluid friction in the heat exchanger. The dimensionless entropy generation number can be obtained by dividing Eq. (28) by $Q/T_{c,o}$ and is taken as the objective function in the optimization process. In this example, the fin height H, the internal span of fins x and the layer number n of the heat exchanger are selected as the design variables and their ranges of values are 3.0–12.0 mm, 1.4–4.2 mm, and 4–30 mm, respectively. The similar optimization process as described in last two examples is implemented. The optimum solution is thus obtained which is documented in Table 10.

From Table 10 it is evident that $\dot{S}_{gen,H}$ and $\dot{S}_{gen,E}$ are the dominant parts in the total entropy generation rate. They are respectively reduced by 44.7 and 44.3 % in comparison with the initial design. Besides, the heat transfer rate increases from 1.829 to 2.132 kW, and the fan power increases from 3.37 to 29.6 W through the optimization process. Note that the irreversibility $\dot{S}_{gen,F}$ related to the fluid friction is quite small in comparison with other irreversibilities.

The amount of power saving through the optimization process can be determined by subtracting the increase of fan power from the increase of heat transfer rate, which is equal to 0.277 kW. Hence the optimization design of the waste heat recovery ventilation system can lead to 1,107 kWh energy saving per year. While the capital investment has not changed, since the heat transfer area is kept constant. Therefore, a notable effect is achieved by the optimization design of the waste heat recovery ventilation system.

The REGN is more advantageous than the EGN in the sense that it avoids the 'entropy generation paradoxes'. Therefore we take the REGN as the objective function and develop an optimization design scheme for the plate-fin type heat exchanger. We find that the optimization process can significantly improve the heat exchanger performance in the sense that the drastic decrease of the power of fans is achieved while the heat load is fixed. Finally the application of this improved entropy generation minimization approach to a waste heat recovery ventilation system results in a notable energy saving.

3 Optimization Design of Shell-and-Tube Heat Exchanger by an Improved Entropy Generation Minimization Approach

In this section based on the EGM a shell-and-tube heat exchanger (STHE) optimization design method is proposed, wherein the dimensionless entropy generation rate obtained by scaling the entropy generation rate on the ratio of heat transfer rate to the inlet temperature of cold fluid is taken as objective function, some geometrical parameters of the STHE are selected as the design variables and the genetic algorithm is employed to solve the optimization design problems. We find that for the case that the heat duty is fixed, not only can the optimization design improve the heat exchanger effectiveness significantly, but also reduce the pumping power drastically.

Table 10 The initial design and optimum result for the waste heat recovery ventilation system

	H (m)	x (m)	n	K_h (W/m^2 K)	Q (kW)	$T_{h,o}$ (K)	$T_{c,o}$ (K)	W (W)	$\dot{S}_{g,H}$ (W/K)	$\dot{S}_{g,T}$ (W/K)	$\dot{S}_{g,F}$ (W/K)	$\dot{S}_{g,E}$ (W/K)	N_{s1}
Initial	0.0112	0.0039	24	13.45	1.829	274.94	286.95	3.37	2.36	0.16	0.0092	2.55	0.744
Final	0.0068	0.0014	30	28.2	2.132	271.9	289.4	29.6	1.305	0.1025	0.0804	1.42	0.365

3.1 Introduction

Energy conservation is one of important measures for keeping sustainable development of world economy. Heat exchangers are used for exchanging thermal energy between two or more media and widely applied to power engineering, petroleum refineries, chemical industries, food industries and so on. The STHE shown in Fig. 10 is the most common heat exchanger type. Therefore to study optimization methods of STHE designs is of great importance for saving energy in heat exchange processes. How to choose the objective function is the key to the success of the optimization of the STHE design. Two groups of objective functions for heat exchanger optimization designs are available: one is based on the first law of thermodynamics and another one is based on the combination of the first and second law of thermodynamics. In recent decades the second group has aroused widespread interest, which includes the entropy generation and exergy. Based on the concept of entropy generation, several heat exchanger performance criteria were proposed [17]. Among them the EGN is the most widely applied one defined by Bejan [12, 13, 18], and the EGM has found wide applications in various thermal engineering problems [13, 29, 30].

From our knowledge the most applications related to the EGM are based on the EGN. However, it was found that the EGN suffers from the 'entropy generation paradoxes' [34, 48]. In order to avoid this paradox the ratio of heat transfer rate to the inlet temperature of the cold fluid can be used for non-dimensionalising the entropy generation rate [34]. The dimensionless entropy generation number defined in such a way is called as the revised entropy generation number in the following discussions and taken as the objective function in the STHE optimization design.

Usually the heat exchanger optimization design with multiple design variables is more practicable, and its global optimum solution is more desirable. The GA is a powerful tool to seeking the global optimum solution of the multiple variable optimization problems in thermal engineering [49–51].

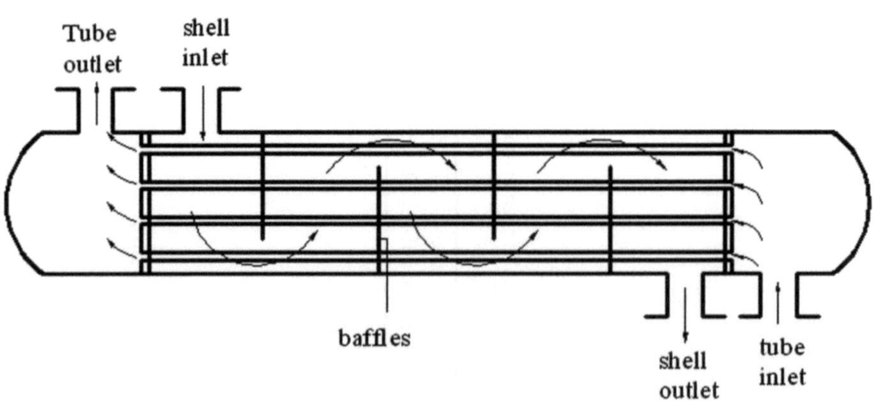

Fig. 10 A typical STHE

In this subsection we attempt to apply the improved EGM to STHE optimization designs. In this method the revised entropy generation number is taken as the objective function, the Bell-Delaware design method is used for the thermal design of STHE, and the GA is employed to solve the resulting optimization problem with multiple design variables. The influence of the selected design variables and the STHE's other parameters on the entropy generation is investigated.

3.2 Optimization Design Method

In the improved EGM for the STHE optimization design developed in the following, the GA will be applied to solve the optimization problems. Its basic principle has been discussed in Sect. 1.2.2. For more details one may refer to Refs. [49–52].

3.2.1 Heat Transfer and Pressure Drop Calculations

Many methods are available for calculating the shell inside diameter of the STHE. The following one is applied in the present work [53]:

$$D_s = (1.1 \cdot \sqrt{n} - 1) \cdot s + 3d_o \tag{29}$$

where D_s is the shell inside diameter, n is the number of tubes, s is the tube pitch, d_o is the tube outside diameter. The tube side heat transfer coefficient U_i is calculated based on the heat transfer correlation [53–55]:

$$U_i = 0.023 \frac{\lambda_1}{d_i} \left(\frac{w_1 d_i}{v_1} \right)^{0.8} Pr_1^{0.3} \tag{30}$$

where λ_1 is the thermal conductivity of the hot fluid, d_i is the tube inside diameter, w_1 is the tube side flow velocity, v_1 is the kinematic viscosity of the hot fluid, Pr_1 is the Prandtl number of the hot fluid.

For the segmentally baffled shell-and-tube exchanger, the Bell-Delaware method is usually used for the STHE thermal design [36, 53, 55]. In this method, the shell-side heat transfer coefficient U_o is determined by correcting the ideal heat transfer coefficient in consideration of the various leakages and bypass flow streams, which is expressed as follows [36, 53]:

$$U_o = j_o G_2 c_{p2} \left(\frac{v_2}{v_{2,w}} \right)^{0.14} Pr_2^{-2/3} \tag{31}$$

where j_o is the shell-side heat transfer factor, G_2 the mass flow velocity of the cold fluid, c_{p2} the specific heat of the cold fluid, v_2 the kinematic viscosity of the cold fluid, $\nu_{2,w}$ the kinematic viscosity of the shell side fluid at wall temperature, Pr_2 the Prandtl number of the cold fluid. The overall heat transfer coefficient U depends on the tube side and shell side heat transfer coefficient and fouling resistances, which is expressed as follows [53]:

$$\frac{1}{U} = \frac{1}{U_i}\left(\frac{d_o}{d_i}\right) + r_t\left(\frac{d_o}{d_i}\right) + \frac{\delta_w}{\lambda_w}\left(\frac{d_o}{d_i}\right) + r_s + \frac{1}{U_o} \tag{32}$$

where r_t is the tube side fouling resistance, δ_w is the tube wall thickness, λ_w is the thermal conductivity of the tube wall, r_s is the shell side fouling resistance.

By neglecting pressure lose in bend for the single tube pass, the total tube side pressure drop ΔP_1 is written as [53]:

$$\Delta P_1 = \left(4f_i\frac{L}{d_i}\left(\frac{\nu_1}{\nu_{1,w}}\right)^{-0.14} + 1.5\right)\frac{\rho_1 w_1^2}{2} \tag{33}$$

where f_i is fanning friction coefficient, L is the tube length, ρ_1 is the density of the hot fluid. In order to calculate the shell side pressure drop, we should first calculate the ideal pressure drops in cross flow section and window section, then correct them by the effective correction factors, and finally sum all the components [36, 53]. The expression of the total pressure drop across the shell side obtained in such a way is [53]:

$$\Delta P_2 = [(N_b - 1)\Delta P_{bk}R_b + N_b\Delta P_{wk}]R_1 + 2\Delta P_{bk}R_b(1 + \frac{N_{cw}}{N_c})R_s \tag{34}$$

More details about the shell-side pressure drop calculation can be found in [36, 53–55]. After the pressure drops are obtained, the total pumping power $\dot{W}$ is written as [46]

$$\dot{W} = \frac{1}{1{,}000\eta}\left(\frac{\dot{m}_1}{\rho_1}\Delta P_1 + \frac{\dot{m}_2}{\rho_2}\Delta P_2\right) \tag{35}$$

where η is the overall pumping efficiency which is set to 80 % in our discussion, $\dot{m}_1$ and $\dot{m}_2$ are respectively the mass flow rates of the hot and cold fluids, ρ_2 is the density of the cold fluid.

3.2.2 The Revised Entropy Generation Number

The irreversibility losses in heat exchanger are quantified in terms of entropy generation. The entropy generation rate caused by heat conduction over finite temperature difference, $\dot{S}_{gen,\Delta T}$, is written as follows [37]:

$$\dot{S}_{\mathrm{gen},\Delta T} = \int\limits_{\mathrm{i}}^{\mathrm{o}} \left(\frac{\dot{m}c_p dT}{T}\right)_{1,2} = \left(\dot{m}c_p\right)_1 \ln\frac{T_1^{\mathrm{o}}}{T_1^{\mathrm{i}}} + \left(\dot{m}c_p\right)_2 \ln\frac{T_2^{\mathrm{o}}}{T_2^{\mathrm{i}}} \tag{36}$$

where T is the temperature, the subscripts 1 and 2 respectively represent the hot and cold fluids, the superscripts i and o stand for the inside and outside. For an incompressible fluid, under nonadiabatic conditions the entropy generation rate related to fluid friction, $\dot{S}_{\mathrm{gen},\Delta P}$, is expressed as [37],

$$\dot{S}_{gen,\Delta P} = \left(\dot{m}\frac{\Delta P}{\rho}\frac{\ln\left(T^{\mathrm{o}}/T^{\mathrm{i}}\right)}{T^{\mathrm{o}} - T^{\mathrm{i}}}\right)_{1,2} = \dot{m}_1 \frac{\Delta P_1}{\rho_1}\frac{\ln\left(T_1^{\mathrm{o}}/T_1^{\mathrm{i}}\right)}{T_1^{\mathrm{o}} - T_1^{\mathrm{i}}} + \dot{m}_2 \frac{\Delta P_2}{\rho_2}\frac{\ln\left(T_2^{\mathrm{o}}/T_2^{\mathrm{i}}\right)}{T_2^{\mathrm{o}} - T_2^{\mathrm{i}}} \tag{37}$$

The total entropy generation rate in heat exchanger is thus written as

$$\begin{aligned}\dot{S}_{\mathrm{gen}} &= \dot{S}_{\mathrm{gen},\Delta T} + \dot{S}_{\mathrm{gen},\Delta P} \\ &= \left(\dot{m}c_p\right)_1 \ln\frac{T_1^{\mathrm{o}}}{T_1^{\mathrm{i}}} + \left(\dot{m}c_p\right)_2 \ln\frac{T_2^{\mathrm{o}}}{T_2^{\mathrm{i}}} + \frac{\Delta P_1}{\rho_1}\dot{m}_1 \frac{\ln\left(T_1^{\mathrm{o}}/T_1^{\mathrm{i}}\right)}{T_1^{\mathrm{o}} - T_1^{\mathrm{i}}} + \frac{\Delta P_2}{\rho_2}\dot{m}_2 \frac{\ln\left(T_2^{\mathrm{o}}/T_2^{\mathrm{i}}\right)}{T_2^{\mathrm{o}} - T_2^{\mathrm{i}}}\end{aligned} \tag{38}$$

When the heat capacity rate of the hot fluid is larger than that of the cold fluid, the outlet temperatures of both fluids can be expressed as

$$T_1^{\mathrm{o}} = T_1^{\mathrm{i}} - \varepsilon\left(T_1^{\mathrm{i}} - T_2^{\mathrm{i}}\right)C^* \tag{39}$$

$$T_2^{\mathrm{o}} = T_2^{\mathrm{i}} + \varepsilon\left(T_1^{\mathrm{i}} - T_2^{\mathrm{i}}\right) \tag{40}$$

where $C^* = (\dot{m}c_p)_{\min}/(\dot{m}c_p)_{\max}$ is the heat capacity rate ratio and obviously $C^* \leq 1$, and for the exchanger effectiveness $\varepsilon$for various flow arrangements one may refer to Refs. [36, 37].

Non-dimensionalising the total entropy generation by the larger heat capacity rate $\left(\dot{m}c_p\right)_{\max}$ leads to the following entropy generation number

$$N_s = \frac{\dot{S}_{gen}}{\left(\dot{m}\cdot c_p\right)_{\max}} \tag{41}$$

which is called the Bejan's definition of EGN in the following discussions. However Bejan's definition of EGN induces the 'entropy generation paradoxes' [34, 48]. In order to avoid this paradox another non-dimensionalisation method for the entropy generation rate was proposed by Hesselgreaves [34],

$$N_{\mathrm{s}1} = \frac{\dot{S}_{\mathrm{gen}} T_2^{\mathrm{i}}}{Q} \tag{42}$$

which is called the revised EGN.

Table 11 Known information of a STHE for example one

	Fluid 1 (tube side)	Fluid 2 (shell side)
Inlet temperature T^{i} (K)	368.15	283.15
Outlet temperature T^{o} (K)	343.15	–
Mass flow rate $\dot{m}$ (kg/s)	50	–
Density ρ (kg/m^3)	970	991.15
Constant pressure specific heat c_p (J/kg K)	4,200	4,174
Kinematic viscosity v (m^2/s)	3.36×10^{-7}	6.96×10^{-7}
Entrance pressure P^{i} (MPa)	6.5	5
Fouling resistance r (m^2 K/W)	0.000086	0.00017
Prandtl number Pr	2.015	4.5878

3.2.3 Optimization of STHE Design

In this subsection, with the revised EGN as the objective function we consider three optimization design problems of the STHE with single tube pass.

Example One

The known information of the STHE under consideration is documented in Table 11. Five design variables are selected and listed as follows:

(a) Tube outer diameter, d_o. The corresponding tube pitch s can be determined based on this parameter [56].
(b) Number of tubes, n. The range of n is from 50 to 500 according to Table 11.
(c) The ratio of the baffle spacing to shell diameter, Bs. Its range is from 20 to 100 % according to the national standard of heat exchanger design in China [56].
(d) The central angle of baffle cut, θ. The value of the baffle cut is recommended between 20 and 45 % of the shell diameter, so the range of θ is from 1.8546 to 2.9413 rad.
(e) The outlet temperature of the cold fluid in the heat exchanger, T_2^{o}. The range of T_2^{o} is from 313.15 to 343.15 K according to Table 11.

The fitness function is defined as follows

$$\text{Fitness} = \max(-N_{s1})$$

Note that the fitness function takes the negative value of the revised EGN.

The constraint conditions are: length-diameter ratio is between 6 and 10; the baffle spacing is greater than 50 mm; the tube and shell side pressure drops are less than 5×10^4 Pa.

The optimization problem described above is solved by the GA with the help of the genetic algorithm toolbox of MATLAB [43]. According to the possible range of the design variables, a random initial population of the design variables is first

Table 12 The optimum results after the evolution of 500 generations for Example One

d_o (m)	n	Bs	θ (rad)	A_o (m^2)	U (W/m^2 K)	T_2^o (K)	$\dot{W}$ (kW)	ε	Ntu
0.016	526	0.70	2.33	99.37	1,320.6	343.11	0.552	0.7054	1.5

created. Then the iterative search process is carried out as shown in Fig. 4 until the stop criteria are satisfied. During this process if any individual does not satisfy one of the constraint conditions, it will be eliminated in the subsequent process of evolution by the penalty function method (set N_s equal to 0.14). The maximum number of generations is set at 500. Initial population includes 20 individuals; floating point coding is adopted. The normalized geometric selection, arithmetic crossover and non-uniform mutation are applied. After 500 generations, the optimal solution of the STHE optimization design problem under consideration is obtained and shown in Table 12.

For each generation in the evolution process, according to the values of the fitness function of every individual, a best individual is selected. The variation of the fitness function with respect to the number of generations is depicted in Fig. 11. Obviously, the fitness value reaches the optimum value when the termination condition is met. The variation of the revised EGN with respect to the heat exchanger effectiveness, the heat capacity rate ratio, the number of exchanger heat transfer units and the total pumping power of the best individual are illustrated in Fig. 12. From Fig. 12a, one can see that as the revised EGN decreases, the heat exchanger effectiveness increases. This is consistent with the thermodynamic theory which states that the undesirable thermodynamic irreversibility quantified by the EGN is detrimental to the thermal performance of the heat exchanger [57]. The heat capacity rate ratio decreases with decreasing the revised EGN as shown in Fig. 12b, which indicates that as the outlet temperature of the cold fluid in the shell side increases and the heat transfer rate keeps constant, the shell side mass

Fig. 11 Variation of the fitness value of the best individual in each generation with respect to the number of generations

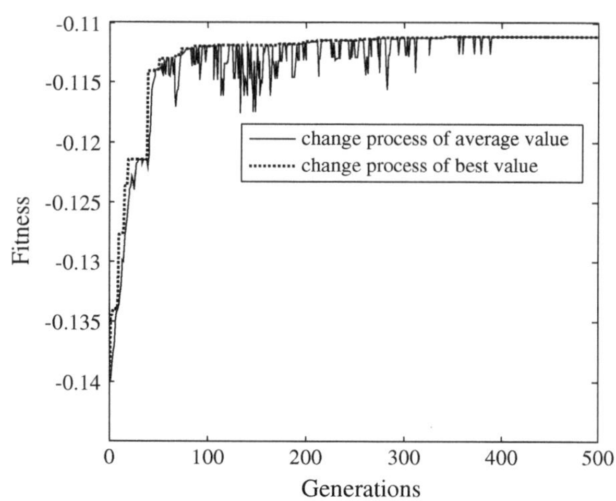

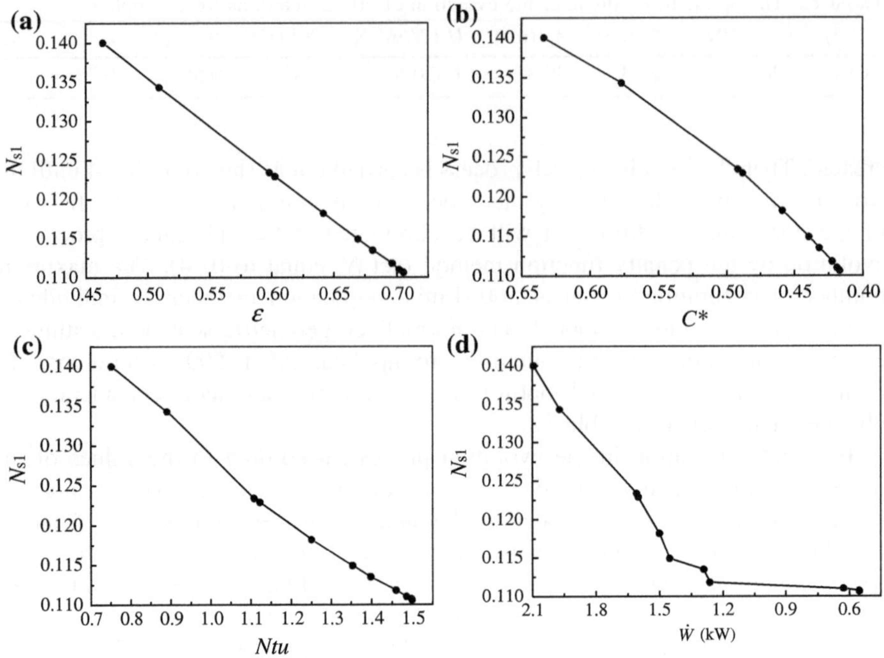

Fig. 12 Variations of the revised EGN with respect to the heat exchanger effectiveness, the heat capacity rate ratio, the number of exchanger heat transfer units and the pumping power. **a** The revised EGN versus the heat exchanger Effectiveness. **b** The revised EGN versus the heat capacity rate ratio. **c** The revised EGN versus the number of exchanger heat transfer units. **d** The revised EGN versus the total pumping power

fluid rate decreases. Therefore if the heat exchanger is used for exchanging heat between the hot and cold waters, the hot water can be saved through this optimization process. From Fig. 12c, one can see that the decrease of the revised EGN leads to the increase of the number of transfer units, this implies that the reduction of the revised EGN is at the cost of increasing heat transfer area when $C_{\min}$ and the overall heat transfer coefficient are fixed. It would be expected that the revised EGN related to the heat conduction over the finite temperature difference should reach the minimum when *Ntu* becomes the infinite, this agrees with the result depicted in Fig. 12c. From Fig. 12d one can see that when the revised EGN declines gradually, the total pumping power generally becomes smaller and smaller due to the decrease of the mass flow rate of cold water.

For this example, if the Bejan's definition of the EGN expressed in Eq. (41) is employed in the definition of the fitness function, it will yield the same result as above. The reason is that where the heat duty is fixed the two types of definitions of the EGN given by Eqs. (41) and (42) are equivalent for the optimization problem under consideration.

Table 13 The known information of a STHE

	Fluid 1 (tube side)	Fluid 2 (shell side)
Inlet temperature T^i (K)	368.15	283.15
Mass flow rate $\dot{m}$ (kg/s)	50	20
Density ρ (kg/m^3)	970	991.15
Constant pressure specific heat c_p (J/kg K)	4,200	4,174
Kinematic viscosity v (m^2/s)	3.36×10^{-7}	6.96×10^{-7}
Entrance pressure P^i (MPa)	6.5	5
Fouling resistance r (m^2 K/W)	0.000086	0.00017
Prandtl number Pr	2.015	4.5878

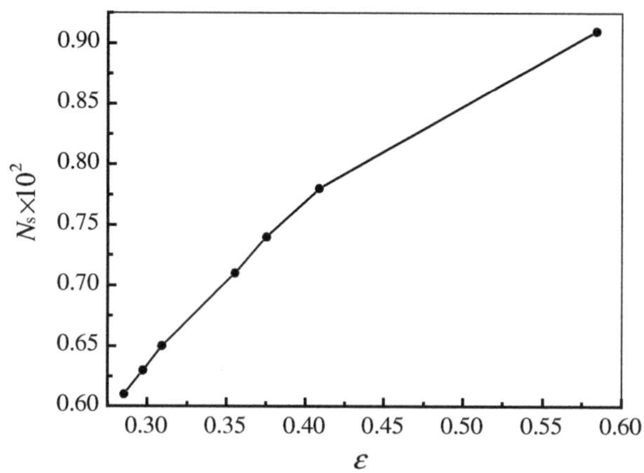

Fig. 13 Variation of the Bejan's EGN with respect to the heat exchanger effectiveness

Example Two

The known data for this example is compiled in Table 13. In this example instead of the outlet temperature of the cold fluid, the tube length is selected as a design variable which ranges from 1 to 12 m [56]. The other design variables are the same as that discussed in Example One.

The EGN defined by Bejan is first taken as the objective function. The GA is applied to solve the optimization problem. For the obtained optimal solution, the relation between the EGN and the exchanger effectiveness is shown in Fig. 13. One can see from this figure that the decrease of EGN leads to the decrease of the heat exchanger effectiveness, which is not the case as suggested by the thermodynamics. This phenomenon is called the 'entropy generation paradox' [34]. Therefore the Bejan's definition of the EGN is not suitable for the present heat exchanger optimization design problem.

If the revised EGN is taken as the objective function, the variations of the revised EGN with respect to the heat exchanger effectiveness and the heat transfer unit are illustrated in Fig. 14a and b, respectively. From Fig. 14a one can see that the decrease of the revised EGN leads to the increase of the heat exchanger

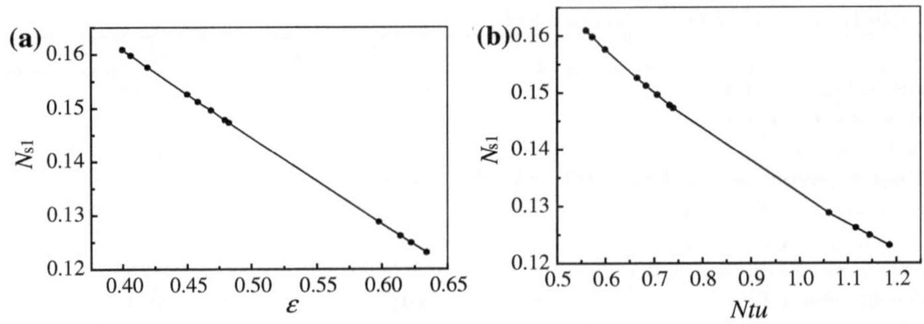

Fig. 14 Variation of the revised EGN with respect to the heat exchanger effectiveness and *Ntu*. **a** The revised EGN versus the heat exchanger effectiveness. **b** The revised EGN versus *Ntu*

effectiveness. Therefore the revised EGN avoids the 'entropy generation paradox', thus demonstrates an obvious advantage over the traditional EGM with the EGN as the objective function.

Example Three

In order to further examine the optimization design of the STHE with the use of the revised EGN and the GA. The following optimization problem is considered. The tube outside diameter (d_o), the number of tubes (n), the ratio of baffle spacing to shell inside diameter (Bs) and the central angle of baffle cut (θ) are selected as the design variables. The heat transfer area is set to 60 m^2. The revised EGN is taken as the objective function for the optimization design of the heat exchanger. After the evolution of 200 generations, the final optimum parameters of the STHE are documented in Table 14. Figure 15 shows a sequence of snapshots of the revised EGN, the heat transfer rate, the heat exchanger effectiveness and the pressure drop of the best individual in the evolution process. From Fig. 15a and b, one can see that the number of heat transfer units increases with the decrease of the revised EGN, so does the heat exchanger effectiveness. Although the variation of the total pumping power with the revised EGN is complicated as shown in Fig. 15c, the overall trend is that the pumping power increases with decreasing the revised EGN.

Note that in this example, the heat transfer area is fixed and the design variables are only related to the geometry of the STHE. Therefore through the optimization design of the geometry of the STHE, the performance of the STHE can be improved significantly.

Table 14 The optimum parameters of the heat exchanger after the evolution of 200 generations

d_o (m)	n	Bs	θ (rad)	K_o (W/m^2 K)	T_1^o (K)	T_2^o (K)	$\dot{W}$ (kW)	ε	*Ntu*
0.016	318	0.36	2.09	1,634.6	346.83	336.7	2.35	0.6308	1.175

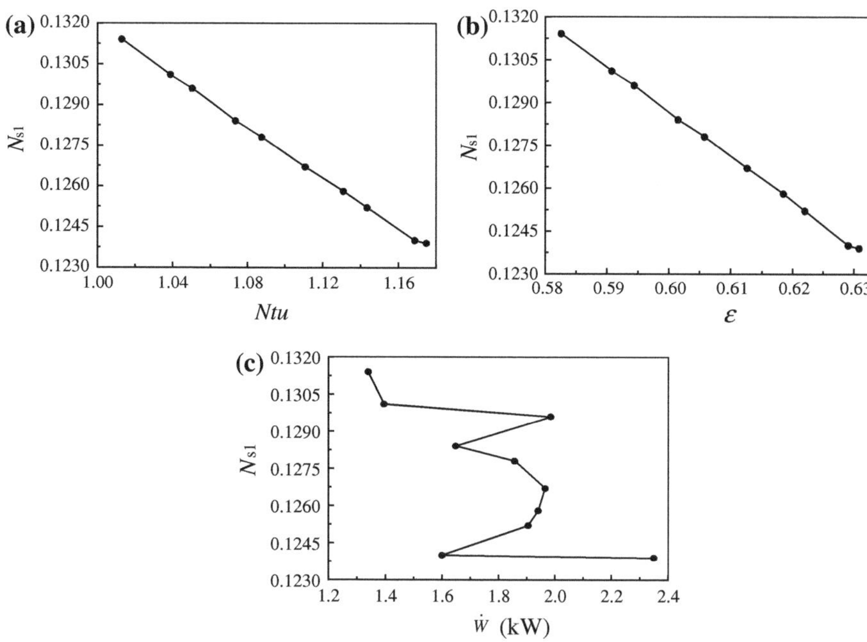

Fig. 15 Variations of the revised EGN with respect to the number of heat transfer units, the heat exchanger effectiveness and the pumping power. **a** The revised EGN versus *Ntu*. **b** The revised EGN versus exchanger effectiveness. **c** The revised EGN versus the pumping power

In summary, based on the second law of thermodynamics an improved entropy generation approach is applied to the optimization design of the STHE. In this approach, the revised EGN which can avoid the entropy generation paradox is taken as the objective function, the GA is applied to solve the multivariable optimization problems, which not only leads to the globe optimum solution but also demonstrates the flexibility to select the design variables and constraint conditions.

Three optimization design examples are presented. In the first example five design variables are selected and the heat load is fixed. It is shown that the optimization design process can remarkably improve the heat transfer effectiveness and save the pumping power.

The second example shows that the EGN defined by Bejan suffers from the 'entropy generation paradox', while the revised EGN can avoid this, in this point the revised EGN demonstrates obvious advantage. Therefore the revised EGN is more suitable for the heat exchanger optimization design.

In the third example, the heat transfer area is fixed. The tube and baffle geometrical parameters are selected as the design variables and thus are allowed to change in the optimization design process. It is found that the optimization design with the revised EGN as the objective function enables us to significantly improve the heat exchanger performance.

In the traditional optimization design of the STHE based on the first law of thermodynamics, the thermodynamic irreversible losses in a heat exchanger has not been accounted for. Furthermore it is found that the traditional optimization design with the total cost as the objective function suffers from decreasing the heat exchanger effectiveness [58]. In this point the optimization approach developed in this section is more advantageous than the traditional heat exchanger optimization design.

4 Multi-objective Optimization Design of Shell-and-Tube Heat Exchanger

In this section a multi-objective optimization of STHE thermal design is developed in the framework of the EGM. The objective of this optimization design is to minimize the dimensionless entropy generation rates related to the heat conduction under finite temperature difference and fluid friction under finite pressure drop. Constraints are specified by the admissible pressure drop and heat exchanger design standards. The genetic algorithm is employed to search the Pareto optimal set of multi-objective optimization problems. We find that the solutions in the Pareto optimal set are trade-off between the pumping power and heat exchanger effectiveness. In some sense, the optimal solution in the Pareto optimal set achieves the largest exchanger effectiveness by consuming the least pumping power under the design requirements and standards. In comparison with the single-objective optimization design, the multi-objective optimization design gives rise to the significant decrease of the pumping power for the same heat exchanger effectiveness.

4.1 Introduction

With the decline of storage capacity of fossil fuels, to use energy sources efficiently is one of most effective ways to reduce the energy demand. Heat exchangers are used for transferring thermal energy between two or more media and widely applied to power engineering, petroleum refineries, chemical industries, food industries and so on. Hence, it is of great importance to improve the performance of heat exchange devices by reducing the unnecessary energy dissipation in the heat exchanger process.

The heat transfer in heat exchangers usually involves the heat conduction under finite temperature difference, the fluid friction under finite pressure drop and fluid mixing. These processes are characterized as irreversible non-equilibrium thermodynamic processes. Inspired by the minimum entropy production principle advanced by Prigogine [59], Bejan [12, 13] proposed the EGM for the heat exchanger optimization design. In this approach, Bejan took into account of two types of the irreversibilities in heat exchanger, namely, the transfer of heat across the stream-to-stream temperature difference and the frictional pressure drop that accompanies the circulation of fluid through the apparatus [12]. Therefore, the total entropy production rate denoted by $\dot{S}_{gen}$ is the sum of entropy productions related to heat conduction and fluid friction. Note that the entropy generation rate usually varies with the mass flow rate, heat transfer rate and other parameters of a heat exchanger. In such a circumstance, it is difficult to assess the performance of heat exchangers only in terms of entropy generation rate. Therefore, it is necessary to non-dimensionalise the entropy generation rate for heat exchanger performance evaluation [17]. Various dimensionless entropy generation rates have been proposed. Among them the most popular one is obtained by non-dimensionalising the entropy generation rate by the capacity flow rate $\dot{m}c_p$, that is, $N_s = \dot{S}_{gen}/\dot{m}c_p$ [16, 60]. The dimensionless entropy generation rate defined in such a way is called the EGN. For a balanced counter-flow heat exchanger, the following approximate expression for N_s was derived by Bejan [12],

$$N_s = \frac{\tau^2}{\frac{4L}{D}St(Re_D, Pr)} + \frac{R}{c_p}\frac{G^2}{\sqrt{2\rho P}}f(Re_D)\frac{4L}{D} \tag{43}$$

where $\tau = |T_{c,i} - T_{h,i}|/\sqrt{T_{h,i}T_{c,i}}$ ($T_{h,i}$ and $T_{c,i}$ are respectively the inlet temperatures of the hot and cold fluids), L is the path length, D the duct hydraulic diameter, Re_D the Reynolds number with the hydraulic diameter as the characteristic length, St the Stanton number which depends on Reynolds number and Prandtl number (Pr), f the friction factor depending on the Reynolds number, c_p the specific heat at constant pressure, P the pressure, ρ the density, G the mass flow rate, R the ideal gas constant. The first term on the right side of Eq. (43) represents the heat conduction irreversibility and the second term accounts for fluid friction irreversibility. It is evident that the duct aspect ratio ($4D/L$) has opposite influence on the heat conduction and fluid friction irreversibilities. Therefore, there exists a best trade-off between the entropy generations caused by heat conduction and fluid friction, respectively. This is the key to the EGM developed by Bejan [12]. However, the entropy generation associated with fluid friction is normally negligible for liquid-to-liquid heat exchangers in comparison with the entropy generation induced by heat conduction, which will be discussed in detail in Sect. 3.2.2. Thus the single-objective optimization of heat exchanger design with the total entropy generation number taken as the single objective function would lead to significant increase of pressure drop. In order to eliminate this drawback of the single-objective optimization of heat exchanger design, we attempt to develop a multi-objective optimization of heat exchanger design, in which the dimensionless entropy generation rates related to heat conduction and fluid friction are set to two separate objective functions. And the multi-objective genetic algorithm is employed to seek the Pareto optimal set which theoretically represent the best trade-off between the entropy generation rates induced by heat conduction and fluid friction, respectively.

Additionally, the method of non-dimensionalising entropy generation rate by the capacity flow rate $\dot{m}c_p$ causes some paradoxes [2, 34]. In an effort to resolve these paradoxes, a number of ways for non-dimensionalising the entropy generation were proposed, such as Q/T_a [20, 35] (Q is the heat transfer rate, T_a is the ambient temperature) and $Q/T_{c,i}$ ($T_{c,i}$ is the inlet temperature of cold fluid in heat exchanger) [34], the latter is employed to non-dimensionalise the entropy generation rate in the following discussion.

4.2 Thermodynamic Optimization of STHE

We consider the optimization of the STHE depicted in Fig. 10. For the heat transfer and pressure drop calculation, please refer to Sects. 2.2.1 and 2.2.2.

4.2.1 The Modified Entropy Generation Number

The irreversible losses in heat exchanger are detrimental to the performance of heat exchanger and can be quantified in terms of entropy generation. By thermodynamic optimization it means to minimize the entropy generation. In heat exchanger the heat conduction under finite temperature difference and fluid friction are two main irreversibilities. In the following, we first calculate the entropy generation rates related to these irreversible losses, and then the optimization design of heat exchanger by the EGM is discussed.

Firstly, the entropy generation rate $\dot{S}_{gen,\Delta T}$ induced by heat transfer process in heat exchanger is expressed as follows [37]:

$$\dot{S}_{gen,\Delta T} = \int_1^0 \left(\frac{\dot{m}c_p dT}{T}\right)_{h,c} = (\dot{m}c_p)_h \ln\frac{T_{h,o}}{T_{h,i}} + (\dot{m}c_p)_c \ln\frac{T_{c,o}}{T_{c,o}} \tag{44}$$

Non-dimensionalising the entropy generation rate by the capacity flow rate $\dot{m}c_p$ yields:

$$N_{s,\Delta T} = \frac{\dot{S}_{gen,\Delta T}}{\dot{m}c_p} \tag{45}$$

Theoretically when $N_{s,\Delta T}$ declines, the irreversible loss in heat exchanger was reduced and the performance of heat exchanger would be improved. While the reality is that for a counter-flow heat exchanger, with increasing the heat exchanger effectiveness, $N_{s,\Delta T}$ first increases and reaches its maximum value, then decreases, as shown in Fig. 16. Therefore, the smaller entropy generation does not mean the better the heat exchanger performance. Bejan called this phenomenon the 'entropy generation paradox' [61]. In order to eliminate this paradox, Hesselgreaves proposed another method for non-dimensionalising entropy generation rate [34],

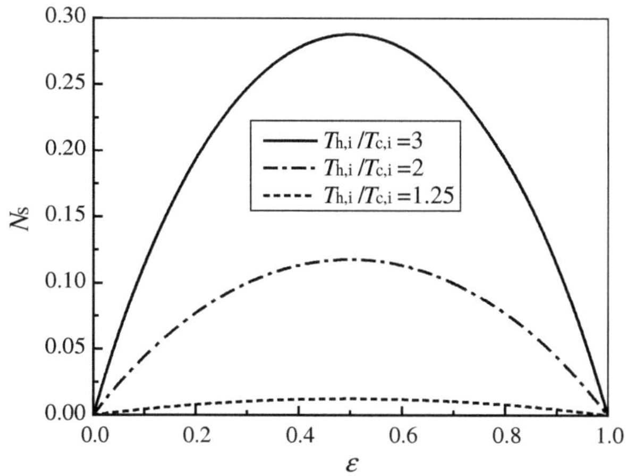

Fig. 16 Bejan's entropy generation number versus the effectiveness

$$N_{s1,\Delta T} = \frac{T_{c,i}\dot{S}_{gen,\Delta T}}{Q} \tag{46}$$

which is called the modified EGN related to the irreversible loss induced by the heat conduction in heat exchanger. $N_{s1,\Delta T}$ decreases monotonously with increasing the exchanger effectiveness as discussed by Hesselgreaves [34]. Therefore, $N_{s1,\Delta T}$ avoids the 'entropy generation paradox'.

For the incompressible fluid and under the non-adiabatic condition the entropy generation rate related to fluid friction is expressed as follows [37]:

$$\begin{aligned}\dot{S}_{gen,\Delta P} &= \left(\dot{m}\frac{\Delta P}{\rho}\frac{\ln(T_o/T_i)}{T_o - T_i}\right)_{h,c} \\ &= \frac{\Delta P_t}{\rho_t}\dot{m}_t\frac{\ln(T_{h,o}/T_{h,i})}{T_{h,o} - T_{h,i}} + \frac{\Delta P_s}{\rho_s}\dot{m}_s\frac{\ln(T_{c,o}/T_{c,i})}{T_{c,o} - T_{c,i}}\end{aligned} \tag{47}$$

For the ideal gas with constant specific heat, the entropy generation induced by flow friction can be written as [57]:

$$\dot{S}_{gen,\Delta P} = -(\dot{m}R)_h \ln\frac{P_{h,o}}{P_{h,i}} - (\dot{m}R)_c \ln\frac{P_{c,o}}{P_{c,i}} \tag{48}$$

where R is the ideal gas constant. Thus, applying the same non-dimensionalising method to the entropy generation related to the flow friction as done to the entropy generation rate caused by heat conduction yields,

$$N_{s1,\Delta P} = \frac{T_{c,i}\dot{S}_{gen,\Delta P}}{Q} \tag{49}$$

Table 15 Tube outer diameter and the corresponding tube pitch

d_o (mm)	10	14	19	20	22	25	30	32	35	38	45	50	55	57
s (mm)	13.4	19	25	26	28	32	38	40	44	48	57	64	70	72

which is the modified EGN associated with the irreversible loss induced by the flow friction.

4.2.2 Single-Objective Optimization

In order to compare with the multi-objective optimization of heat exchanger design, we first discuss the single-objective optimization of STHE design. The total modified EGN N_{s1} is,

$$N_{s1} = N_{s1,\Delta T} + N_{s1,\Delta P} \tag{50}$$

which is taken as the objective function. The known data for the heat exchanger design are documented in Table 13. The working fluids in tube and shell sides are water in our consideration. The design variables and their ranges are selected as follows:

1. The tube outer diameter, d_o, its discrete values and the corresponding tube pitches are listed in Table 15.
2. The number of heat exchange tubes, n, ranging from 50 to 550;
3. The ratio of the baffle spacing to the shell inner diameter, Bs, varies between 0.2 and 1.0;
4. The central angle of baffle cut, θ, ranging from 1.8546 to 2.9413 in radian.

The constraint conditions for the heat exchanger design are:

1. Length-diameter ratio is between 6 and 10;
2. The baffle spacing is greater than 50 mm;
3. The tube side pressure drop is less than 5×10^4 Pa;
4. The shell side pressure drop is less than 5×10^4 Pa [56].

In our design the heat transfer surface area is fixed at 60 m^2. This optimization problem formulated above will be solved by the GA. For each generation in the evolution process, according to the values of the fitness function of every individual, a best individual is selected. The variations of the corresponding values of $N_{s1,\Delta T}$, $N_{s1,\Delta p}$ and N_{s1} with respect to the number of generations are depicted in Figs. 17 and 18.

From Fig. 17 one can see that in comparison with the initial value, the modified EGN decreases by 10.44 % through the optimization process. It seems that the single-objective optimization of heat exchanger works well. However, the observation of Fig. 18 shows that although the modified EGN related to heat conduction is reduced by 10.60 % through the optimization process, the modified EGN related to flow friction is enlarged by 324 %. Why the flow friction has such a significant increase? Bejan introduced the irreversibility distribution ratio ϕ to describe the

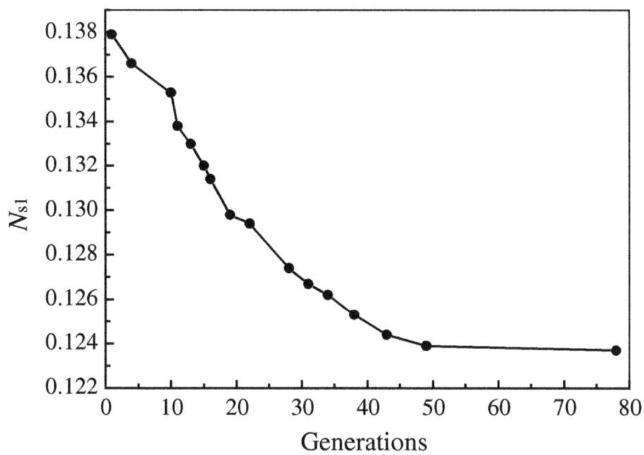

Fig. 17 Variation of N_{s1} with the number of generations

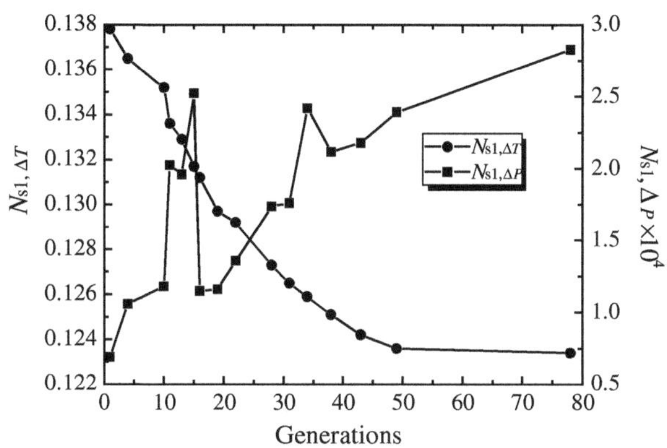

Fig. 18 Variations of $N_{s1,\Delta T}$ and $N_{s1,\Delta P}$ with the number of generations for single-objective optimization

relative importance of the two irreversibilities related to the heat conduction and flow friction in heat exchanger [17]. According to Eqs. (44) and (47), the irreversibility distribution ratio for incompressible fluid in one passage of a heat exchanger is given as follows,

$$\phi = \frac{\dot{S}_{gen,\Delta T}}{\dot{S}_{gen,\Delta P}} = \frac{c_p \rho (T_o - T_i)}{\Delta P} \tag{51}$$

Particularly, for the above single-objective optimization design of the heat exchanger with the known data listed in Table 13, the irreversibility distribution ratio in the shell-side is,

$$\phi = \frac{\dot{S}_{gen,\Delta T}}{\dot{S}_{gen,\Delta P}} = \frac{c_p \rho (T_o - T_i)}{\Delta P} = 4{,}481.8 \tag{52}$$

It is evident that the entropy generation rate related to heat conduction significantly overweight that caused by flow friction. If the working fluid is water, ϕ can be estimated in the following manner. According to Ref. [62] we have,

$$c_p \sim 10^3, \ \rho \sim 10^2 \tag{53}$$

The order of magnitude for the difference between outlet temperature and inlet temperature is about 10. The upper limit of pressure drop in shell-and-tube heat exchanger is about 5×10^4 Pa [56]. Therefore, the irreversibility distribution ratio ϕ is usually no less than 10^2 in most situations. For the other liquids, the order of magnitude of the irreversibility distribution ratio can also be obtained by using magnitude analysis according to their thermodynamic properties [62]. In order to correctly reflect the influence of flow friction in the EGM, one may form a single objective function by linearly combing separate objectives using some fixed weights prescribed in [63]. The objective function expressed in Eq. (50) can be regarded as the sum of $N_{s,\Delta T}$ and $N_{s,\Delta P}$ which are multiplied by weight coefficient 1. The weight coefficient might be 10, 100 or 1,000, depending on the user's will. This method seems to be very straightforward, but different choices of the weight usually give rise to different solutions for a same optimization problem [63], and the weight coefficients have no definite physical meaning. Furthermore, setting an appropriate balance between the objectives to result in a full exploration of the desired range of trade-offs within the objective space, can be difficult, especially where a non-convex objective space is explored [64].

Another alternative to improving the single-objective optimization design of heat exchanger in the framework of the EGM is to utilize the multi-objective optimization approach. Specifically, the entropy generation rates induced by the heat conduction and flow friction can be set to two separate objective functions. Then the multi-objective genetic algorithm is employed to solve this multi-objective optimization problem. Theoretically, it leads to a set of solutions that represent the best trade-off between the two objective functions [65]. In addition, the GA demonstrates obvious advantages over other methods for solving optimization problems. Firstly it does not necessitate the calculation of the objective function gradient with respect to the design variables, which is particularly helpful for multi-objective optimization. Secondly the GA usually explores a large portion of the design space and converges to the global optimum. Therefore, we attempt to establish the multi-objective optimization of heat exchanger design based on the EGM in the following subsection.

4.2.3 Multi-objective Optimization Design

Mathematically, the multi-objective optimization simultaneously minimizes several objectives, with a number of inequality or equality constraints. It can be mathematically formulated as follows:

$$\min_{\mathbf{x}\in X} \mathbf{f}(\mathbf{x}) = [f_1(\mathbf{x}), f_2(\mathbf{x}), \ldots, f_k(\mathbf{x})] \tag{54}$$

Subject to:

$$g_j(\boldsymbol{x}) = 0, \quad j = 1, 2, \ldots, M$$
$$h_k(\boldsymbol{x}) \leq 0, \quad k = 1, 2, \ldots, K$$

where x is a vector and is also called the decision vector, X is the parameter space. If and only if, $f_i(\mathbf{x}) \leq f_i(\mathbf{y})$ for $i = 1, 2, \ldots k$ and $f_j(\mathbf{x}) < f_j(\mathbf{y})$ for least one objective function j, a feasible solution $\mathbf{x}$ is said to dominate another feasible solution $\mathbf{y}$. A solution which is not dominated by any other solution in the feasible region is called Pareto optimal solution. The set of all non-dominated solutions in X is called as the Pareto optimal set P^*, the values of objective functions corresponding to the Pareto optimal set are called Pareto front (denoted as PF^*) [65, 66],

$$PF^* := \{\mathbf{f}(\mathbf{x}) | \mathbf{x} \in P^*\} \tag{55}$$

Specifically, for the multi-objective optimization design of heat exchanger the modified entropy generation numbers related to heat conduction and flow friction, respectively, are taken as two separate objective functions. The design parameters, their bounds and the constraints remain the same as specified in Sect. 3.2.2 for the single-objective optimization design. For exemplifying the multi-objective optimization approach, only four design variables are selected there. The optimization problems with more design variables can be solved by following the similar procedure.

In addition, the population type is *double vector*, and the size of initial population is set as 75. *Feasible population operator* is selected to create a well-dispersed initial population that satisfies all bounds and constraints of design parameters. The algorithm for choosing parents from individuals is *tournament*. Set *crossover fraction* to 0.8 so that 80 % of the next generation is produced by *crossover operator*. *Scattered crossover operator* is used to combine two individuals, or parents, to form a child for the next generation. *Gaussian Mutation operator* is used, the values of *scale* and *shrink* parameters are fixed as one, which means that the standard deviation shrinks linearly from 1 to 0 as the last generation is reached. The direction of *migration* is forward, i.e., the nth subpopulation migrates into the $(n + 1)$th subpopulation, and *migration* takes place every 20 generations.

A controlled elitist genetic algorithm (a variant of NSGA-II [65]) is adopted for searching the optimal solution, which can help increase the diversity of the population even if they have lower fitness values. The diversity of population is controlled by the elite members of the population in the process; the *distance-crowding* function helps to maintain diversity by favoring individuals that are relatively far away on the front. The *Pareto fraction* is set to 0.35 so as to limit the number of individuals in the current population that are on the *Pareto front* to 35 % of the population size. The total number of generations is set to 1,000 which serves as the stopping criteria to terminate the iterative process.

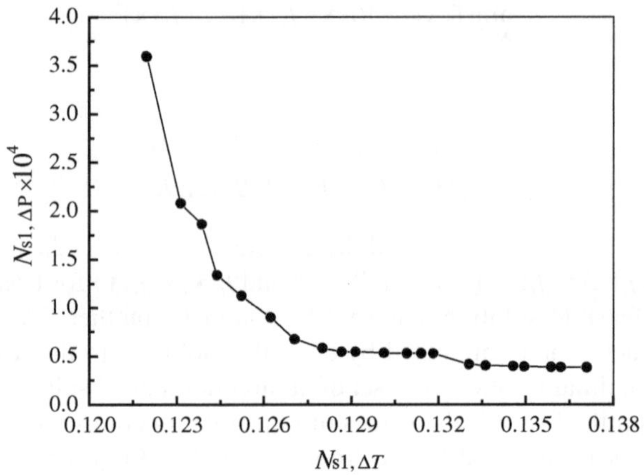

Fig. 19 The Pareto front obtained by multi-objective optimization of STHE design with the fixed heat transfer area

Multi-objective Optimization Design of STHE with Fixed Heat Transfer Area

In this subsection, the multi-objective optimization method is applied to design a STHE with the fixed heat transfer area. The Pareto front obtained by the multi-objective optimization design is shown in Fig. 19. At the leftmost point of the curve representing the Pareto set, $N_{s1,\Delta T}$ reaches its minimum value, while $N_{s1,\Delta P}$ achieves its maximum value. Similarly, on the rightmost point $N_{s1,\Delta T}$ reaches its maximum value and $N_{s1,\Delta P}$ arrives at its minimum value. It indicates that the multi-objective optimization solutions are the best trade-off between the two irreversible losses caused by heat conduction and flow friction, respectively. Therefore, for the multi-objective optimization the entropy generation caused by flow friction is not neglected, but plays an important role in determining the optimal solutions.

In order to illustrate the meaning of the Pareto optimal set, the pumping power and heat exchanger effectiveness corresponding to the individual solution in the Pareto optimal set are depicted in Fig. 20. One can see that the total region is divided into two regions by the Pareto optimal set. The designs with the pumping power and exchanger effectiveness located in the region I are feasible, but not optimal. The ones in the region II are infeasible. Therefore, the solution in the Pareto set is optimal in the sense that its exchanger effectiveness achieves the maximum value by consuming the least pumping power under the design requirements and constraints. Note that normally the optimal solution for the multi-objective optimization design is not unique. The designer can select one from the Pareto set according to the specific design requirements. Therefore, the multi-objective optimization design of heat exchanger is more flexible than the single-objective optimization design.

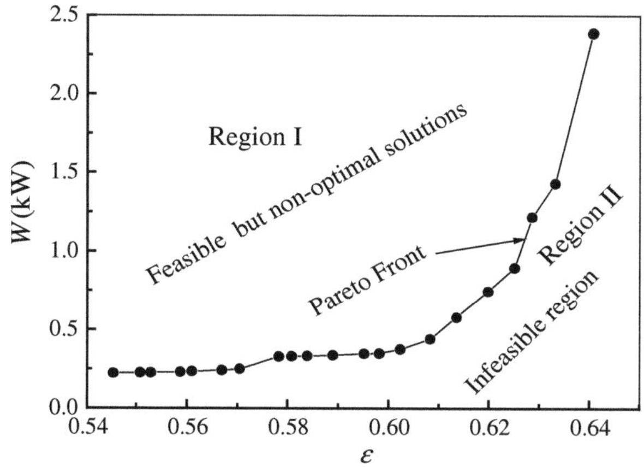

Fig. 20 The pumping power and exchanger effectiveness for Pareto optimal set

Multi-objective Optimization Design of STHE with Fixed Heat Load

In this subsection, the multi-objective optimization design of heat exchanger is applied to the case that the heat load is given. The known data for the heat exchanger design is documented in Table 11. In comparison with the last example, the outlet temperature of cold fluid is selected as an additional design variable. Its values range from 313.15 to 343.15 K. The Pareto front obtained by the multi-objective optimization is shown in Fig. 21. The randomly selected initial design solutions satisfying the design requirements and constraints, and some representative optimal solutions in the Pareto set are listed in Table 16. From this table, one

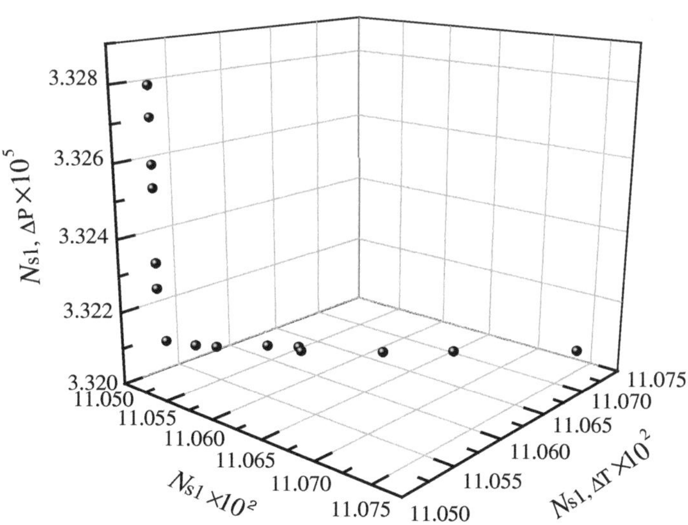

Fig. 21 The relations of N_{s1} with $N_{s1,\Delta T}$ and $N_{s1,\Delta P}$ in the Pareto optimal set for fixed heat duty

Table 16 The initial designs and the optimal solutions for the heat exchanger with the fixed heat duty

	d_o (m)	n	Bs	θ(rad)	$T_{c,o}$ (K)	NTU	W (W)	ε	$Ns1, \Delta T$	$Ns1, \Delta P$	N_{s1}
Initial design	0.016	361	0.72	2.81	314.54	0.55	1,555	0.37	0.152	2.11×10^{-4}	0.152
	0.016	327	0.68	2.65	322.61	0.75	2,094	0.46	0.140	2.77×10^{-4}	0.140
	0.016	327	0.68	2.35	326.47	0.89	1,970	0.51	0.134	2.59×10^{-4}	0.134
Multi-objective	0.022	283	1.000	2.2675	343.13	1.5000	260.35	0.7100	0.1105	3.320×10^{-5}	0.1105
	0.022	283	0.995	2.2662	343.15	1.5000	260.84	0.7100	0.1105	3.332×10^{-5}	0.1105
	0.022	283	1.000	2.6666	343.15	1.5008	260.79	0.7059	0.1105	3.327×10^{-5}	0.1105
	0.022	283	1.000	2.2836	342.99	1.4928	260.29	0.7040	0.1107	3.321×10^{-5}	0.1107
	0.022	283	0.998	2.2656	343.08	1.5008	260.52	0.7059	0.1105	3.324×10^{-5}	0.1105

Table 17 The optimal solution of the single-objective optimization design for heat exchanger with the fixed heat duty

d_o (m)	n	Bs	θ (rad)	$T_{c,o}$ (K)	K_o (W/m^2 K)	NTU	W (W)	ε	$Ns1, \Delta T$	$Ns1, \Delta P$	N_{s1}
0.016	526	0.70	2.33	343.08	1,320	1.5	552	0.71	0.111	6.98×10^{-5}	0.11

can see that in compassion with the initial designs, the exchanger effectiveness of the optimal solutions is significantly improved and the pumping power is drastically reduced, but at the cost of the increase of the number of exchanger heat transfer units. In comparison with the result of the single-objective optimization shown in Table 17, both approaches yield almost the same heat exchanger effectiveness which is about 0.71. However, the pumping power required for the single-objective optimization as shown in Table 17 is 552 W, while for the multi-objective optimization approach it is only about 260 W as shown in Table 16. Therefore, more than 50 % decrease of the pumping power has been achieved for the multi-objective optimization in comparison with the single-objective optimization approach. As expected, the multi-objective optimization approach avoids the drawback of the single-objective optimization approach which has not correctly reflected the influence of the fluid friction in the EGM. Therefore, the multi-objective optimization design of heat exchanger, which can give rise to the best trade-off between the dimensionless entropy generation rates caused by heat conduction and flow friction, respectively, is more suitable for the heat exchanger optimization design.

The single-objective optimization of heat exchanger design based on the EGM suffers the drawback that the role played by the entropy generation related to the flow friction has not sufficiently been taken into account for the fluid-to-fluid heat exchanger. In an attempt to resolve this problem, a multi-objective optimization approach for heat exchanger design is developed, in which the entropy generations induced by heat conduction and flow friction are taken as two separate objective functions. Theoretically the optimal solutions in the Pareto set obtained by the multi-objective genetic algorithm are the best trade-off between the two irreversible losses due to heat transfer and flow friction. The results for two heat exchanger optimization design problems show that for achieving the same exchanger effectiveness the multi-objective optimization of heat exchanger design requires less pumping power in comparison with the single-objective optimization design. Therefore, the multi-objective optimization is more suitable for implementing EGM in the heat exchanger design.

The multi-objective optimization yields the Pareto optimal set which includes several optimal solutions. This gives the designer a great flexibility to choose a final solution from the Pareto optimal set by fully considering the specific requirements and constraints of the heat exchanger design. Furthermore, in the multi-objective optimization of heat exchanger design, all kinds of requirements, such as cost, weight and so on, may be treated as the optimization constraints and can be mathematically formulated in the multi-objective optimization of heat exchanger design. Therefore, the multi-objective optimization of heat exchanger design demonstrates obvious advantages and great flexibility.

5 The Second Law of Thermodynamics in Terms of Entransy

In this section the thermodynamic basis of the concepts of entransy and entransy dissipation is presented [4]. We first define the entransy and entransy dissipation in the Carnot cycle. The Carnot's theorem is restated in terms of entransy. Then the results are extended to the arbitrary cycles. The second law of thermodynamics is thus re-described by the entransy and entransy dissipation.

5.1 Introduction

Historically, the study of the efficiency of heat engine led to the second law of thermodynamics which places constraints upon the direction of heat transfer and the attainable efficiencies of heat engines [67]. The concept of entropy introduced later has stretched the second law of thermodynamics across almost every discipline of science. The second law analysis has found widespread applications in various engineering problems [61, 68–73]. Although there has been effort to modify the expression of the second law of thermodynamics [74–78] and to improve the classical thermodynamics by considering the Carnot construction cycling in a finite time [79–81], the eminent position of entropy in thermodynamics has not been questioned. However, the definition of entropy in the classical thermodynamics exhibits abstract and ambiguous nature, which was noted even by Clausius who first defined the entropy in the classical thermodynamics [82]. Furthermore during the applications of the concepts of entropy and entropy generation in thermal engineering, some paradoxes and inconsistence have been found. Herwig found that for heat transfer in rough pipes there exists a big difference between the optimization results obtained from the entropy generation minimization and the performance evaluation criterion based on the heat transfer theory [83]. Bertola and Cafaro found that the principle of minimum entropy generation is not compatible with continuum mechanics [84]. By analyzing the heat exchanger effectiveness and entropy generation of eighteen types of heat exchangers Shah and Skiepko found that the heat exchanger effectiveness can be maximum, minimum or in between when the entropy generation achieves its minimum value [85]. We also found that the dimensionless entropy generation defined by Bejan [57] is not consistent with the exchanger effectiveness which describes the heat exchanger performance [8].

As an attempt to resolve the 'entropy generation paradox' induced by the entropy generation number Guo et al. defined two new physical concepts called entransy and entransy dissipation for describing the heat transfer ability and irreversibility of heat transfer, respectively [86]. Based on the concept of entransy dissipation, an equivalent thermal resistance of heat exchanger was defined and the relationship between the exchanger effectiveness and the thermal resistance was established [87]. By introducing a new approach to non-dimensionalising the entransy dissipation in heat exchanger we defined an entransy dissipation number which can serve as the heat exchanger performance evaluation criterion [8]. Liu et al. developed a minimum entransy dissipation principle for the optimization of transport networks [88]. Chen et al. applied the entransy dissipation extremum principle to optimize the convective heat transfer processes and made a comparison with the minimum entropy generation principle [89]. In Ref. [90], the entransy dissipation theory was extended to deal with the coupled heat and mass transfer processes in evaporative cooling systems. In Ref. [91], Cheng and Liang defined the entransy flux and entransy function for the thermal radiation in enclosures with opaque surfaces, and developed the minimum principle of radiative entransy. Chen et al. proposed an entransy dissipation rate minimization approach for the disc cooling system and investigated the influence of various system parameters on the entransy dissipation rate [92].

Although the concepts of entransy and entransy dissipation have been successfully applied to analyze the heat transfer processes and demonstrate some advantages over the entropy and entropy generation, how to define these concepts from the thermodynamic point of view is still an open question. In this subsection, we attempt to seek the thermodynamic basis of the entransy dissipation theory.

5.2 Carnot's Theorem in Terms of Entransy

We start with the Carnot cycle. In this cycle, the heat engine absorbs heat Q_1 from the hot reservoir at the temperature T_1 (absolute temperature is always assumed in the following discussion), converts part of heat to work W and discards the rest of the heat to the cold reservoir at the temperature T_2. For this cycle, Carnot's theorem states that [67]

$$Q_1/T_1 \leq Q_2/T_2 \tag{56}$$

where the equality and inequality signs correspond to the reversible and irreversible heat engines, respectively. The efficiency of a reversible engine is defined as

$$\eta = 1 - \frac{T_2}{T_1} \tag{57}$$

Carnot's theorem dictates that reversible engines have the maximum efficiency [67].

Equivalently, Inequality (56) can be rewritten as follows

$$Q_1(T_1 - T_2) \geq (Q_1 - Q_2)T_1 \tag{58}$$

Inspired by this Inequality, we define $G = Q_1(T_1 - T_2)$ as the entransy gained by the heat engine from the hot reservoir in the Carnot cycle. From this definition, one can see that the larger the amount of heat Q_1 and the temperature difference between the hot and cold reservoir, the greater the entransy obtained by the heat engine in the Carnot cycle. Obviously, the larger entransy means higher ability for heat engine to perform work. Guo et al. showed that the entransy can describe the heat conduction ability [86]. Therefore, we may say that the entransy defined here quantifies the energy transfer ability including the ability to deliver work and transfer heat.

To obtain the efficiency of a reversible heat engine, firstly we know that the law of conservation of energy dictates that $W = Q_1 - Q_2$. We define $G_W = (Q_1 - Q_2)$ $T_1 = WT_1$ and regard it as the entransy consumed by the heat engine for delivering work W to the system's exterior. For the reversible heat engine, Inequality (58) reduces to:

$$Q_1(T_1 - T_2) = (Q_1 - Q_2)T_1 = WT_1 \tag{59}$$

which indicates that the entransy gained by the reversible heat engine in the Carnot's cycle is completely converted to deliver the work W. While for the irreversible heat engine, Inequality (58) becomes

$$Q_1(T_1 - T_2) > (Q_1 - Q_2)T_1 \tag{60}$$

which shows that only part of the entransy obtained by the heat engine in the Carnot's cycle is used for performing the work W, the rest is consumed by the irreversibility occurring in the heat engine. In order to quantify the irreversibility in the heat engine, we define the entransy generation in parallel with the entropy generation in the following way:

$$G_g = (Q_1 - Q_2)T_1 - Q_1(T_1 - T_2) \tag{61}$$

While $G_{diss} = -G_g$ is called the entransy dissipation which represents the entransy consumed by the irreversibility in the heat engine. Since the reversible heat engine converts the total entransy gained during the Carnot cycle to perform work, we may define the efficiency of the heat engine in terms of entransy as follows

$$\eta_E = \frac{WT_1}{Q_1(T_1 - T_2)} = \frac{G_W}{G} \tag{62}$$

Notice that the reversible heat engine achieves the maximum value of η_E which is equal to 100 %. This is the statement of Carnot's theorem in terms of entransy.

5.3 The Second Law of Thermodynamics

Carnot's theorem has played a pivotal role on the development of the classical thermodynamics. By generalizing Inequality (56) to an arbitrary cycle, Clausius introduced the concept of entropy which is a physical quantity as fundamental and universal as energy. In this section, we attempt to generalize Inequality (58) to an arbitrary cycle. Firstly, Inequality (58) can be rewritten as follows:

$$Q_1 \Delta T + T_1 \Delta Q \geq 0 \tag{63}$$

where $\Delta T = T_1 - T_2$, $\Delta Q = -(Q_1 - Q_2)$. Note that since the system delivers energy to its exterior, therefore, $\Delta Q < 0$.

An arbitrary cycle can be decomposed into a group of Carnot's cycles denoted as $C_i \, (i = 1, 2, \ldots n)$ [67]. Applying Inequality (63) to the i-th Carnot's cycle yields:

$$Q_{a,i} \Delta T_i + \Delta Q_i T_{h,i} \geq 0 \tag{64}$$

where $Q_{a,i}$ is the heat absorbed from the hot reservoir at temperature $T_{h,i}$, $\Delta Q_i = -(Q_{a,i} - Q_{r,i})$ ($Q_{r,i}$ is the heat discarded into the cold reservoir), $\Delta T_i = T_{h,i} - T_{c,i}(T_{c,i}$ is the temperature of the cold reservoir). If the number of Carnot's cycles under consideration tends to the infinity, the temperature difference between the hot and cold reservoirs approaches to infinitesimal, Inequality (64) becomes

$$\oint QdT + T\delta Q \geq 0 \tag{65}$$

For a reversible cycle, Inequality (65) reduces to

$$\oint QdT + T\delta Q = 0 \tag{66}$$

This equation suggests us to define a quantity E called the entransy as follows

$$dG = QdT + T\delta Q \tag{67}$$

which only depends on the initial and final states of a reversible process. Thus the entransy is a state variable. If G_A and G_B are values of this variable in the initial state A and final state B, respectively, we have

$$G_B - G_A = \int_A^B QdT + T\delta Q \tag{68}$$

If the temperature remains fixed, it follows from Eq. (68) that for a reversible flow of heat δQ, the change in entransy is $T\delta Q$, while the change in entropy is $\delta Q/T$. For this case we get the following relationship between the entransy and entropy,

$$dG = T^2 dS \tag{69}$$

Consider an irreversible process I which starts from the equilibrium state A and ends at the equilibrium state B. In order to form a cycle, we add a reversible process R from the state B to state A. Then from Inequality (65), we have

$$\int_{A_I}^{B_I} QdT + T\delta Q > \int_{A_R}^{B_R} QdT + T\delta Q \tag{70}$$

The subscripts R and I represent the reversible and irreversible processes, respectively. The application of Eq. (68) on Inequality (70) yields

$$\int_{A_I}^{B_I} QdT + T\delta Q > G_B - G_A \tag{71}$$

Thus in parallel with the entropy generation we can define the entransy generation G_g of the irreversible process as follows

$$G_g = (G_B - G_A) - \int_{A_I}^{B_I} QdT + T\delta Q < 0 \tag{72}$$

Note that for the reversible process $G_g = 0$. We define $G_{diss} = -G_g$ as the entransy dissipation. Therefore, the entransy dissipation quantifies the entransy consumed by the irreversibility in the irreversible process. For an irreversible process, the second law of thermodynamics states that the entransy generation is always negative or the irreversible process always decreases the system's ability to do work and transfer heat. From the entransy's definition (68), it is evident that the entransy is an extensive quantity. Subsequently, Inequality (72) is also valid under the local equilibrium assumption.

We may express the system's change in entransy as a sum of two parts

$$dG = d_eG + d_iG, \quad \text{and} \quad d_iG \leq 0 \tag{73}$$

In which d_eG is the entransy change due to exchange of matter and energy with the exterior of the system and d_iG is the entransy generation produced by the irreversible processes occurring in the system. For the closed systems that exchange energy, but not matter, we have

$$d_eG = QdT + T\delta Q \quad \text{and} \quad d_iG \leq 0 \tag{74}$$

For open systems that exchange both matter and energy:

$$d_eG = QdT + T\delta Q + (d_eG)_{matter} \quad \text{and} \quad d_iG \leq 0 \tag{75}$$

Finally, for the isolated systems, we have

$$d_e G = 0, \quad \text{and} \quad d_i G \leq 0 \tag{76}$$

Therefore, in the isolated systems the entransy never increases, namely the energy transfer ability can not increase.

5.4 Entransy Dissipation Due to Heat Conduction

Now let us consider the heat conduction process discussed by Kondepudi and Prigogine [67]. The system under consideration is an isolated system and consists of two parts, each having a well-defined temperature. Let the temperatures of two parts be T_1 and $T_2(T_1 > T_2)$, respectively. δQ is the amount of heat flow from the hotter part to colder part in a time period dt. Since this isolated system does not exchange entransy with its exterior, $d_e G = 0$. Assume the volume of each part is constant, thus $dW = 0$. Hence the energy change in each part is solely due to the flow of heat: $dU_i = \delta Q_i \ (i = 1, 2)$. In accordance with the first law, the heat gained by one part is equal to the heat lost by the other. Therefore, $-\delta Q_1 = \delta Q_2 = \delta Q$. The total change in entransy $d_i G$ of the system is the sum of the changes of entransy in each part due to the flow of heat

$$d_i G = -T_1 \delta Q + T_2 \delta Q = -(T_1 - T_2)\delta Q \tag{77}$$

In terms of the rate of flow of heat $\delta Q/\delta t$, the rate of entransy generation can be written as follows

$$\frac{d_i G}{dt} = -(T_1 - T_2)\frac{\delta Q}{\delta t} \tag{78}$$

The rate of heat flow $J_Q = \delta Q/\delta t$ is given by the Fourier law of heat conduction $J_Q = k(T_1 - T_2)$, in which k is the coefficient of heat conductivity, therefore

$$\frac{d_i G}{dt} = -k(T_1 - T_2)^2 \leq 0 \tag{79}$$

Accordingly, the rate of entransy dissipation is written as

$$\frac{d_i G_{diss}}{dt} = k(T_1 - T_2)^2 \geq 0 \tag{80}$$

This equation is consistent with the expression of the entransy dissipation function obtained by Guo et al. from the Fourier heat conduction equation [86]. Due to the flow of heat from the hot part to cold part, the temperatures of both parts eventually become equal, and the entransy dissipation rate tends to zero and the system reaches the equilibrium state. Therefore, the entransy dissipation rate must vanish at the state of equilibrium, which is called as the principle of minimum entransy dissipation rate. This is the counterpart of the principle of the minimum entropy production. Generally, for three-dimensional steady heat conduction without heat source, the entransy dissipation rate is expressed as

$$\frac{d_i G_{diss}}{dt} = \int_V \frac{1}{2} k (\nabla T)^2 dV \tag{81}$$

where V is the volume of the heat conduction medium. The principle of minimum entransy dissipation rate is mathematically formulated as

$$\delta \int_V \frac{1}{2} k (\nabla T)^2 dV = 0 \tag{82}$$

which is consistent with the least entransy dissipation principle established by Guo et al. [86]. Equation (82) is equivalent to the following Euler–Lagrange equation:

$$\frac{\partial}{\partial x}\left(\frac{\partial F}{\partial\left(\frac{\partial T}{\partial x}\right)}\right) + \frac{\partial}{\partial y}\left(\frac{\partial F}{\partial\left(\frac{\partial T}{\partial y}\right)}\right) + \frac{\partial}{\partial z}\left(\frac{\partial F}{\partial\left(\frac{\partial T}{\partial z}\right)}\right) = 0 \tag{83}$$

with

$$F = \frac{1}{2} k (\nabla T)^2 \tag{84}$$

Substituting Eq. (84) into Eq. (83) yields

$$\frac{\partial^2 T}{\partial x^2} + \frac{\partial^2 T}{\partial y^2} + \frac{\partial^2 T}{\partial z^2} = 0 \tag{85}$$

which is exactly the governing equation of the steady heat conduction without heat sources based on the Fourier law. While Bertola and Cafaro found that the principle of minimum entropy production can not lead to the governing equation of the steady Fourier heat conduction [84]. In this point, the principle of minimum entransy dissipation rate demonstrates an obvious advantage.

5.5 Closed Systems

In this section, the thermodynamic analysis of a closed system in terms of entransy and entransy dissipation is presented. This system has been analyzed by Bejan in terms of entropy generation [13]. The closed system under consideration is illustrated in Fig. 22. In this figure we follow the convention that the work transfer out of the system and heat transfer into the system are set to be positive. For any process between equilibrium states 1 and 2, the first law of thermodynamics is formulated as follows

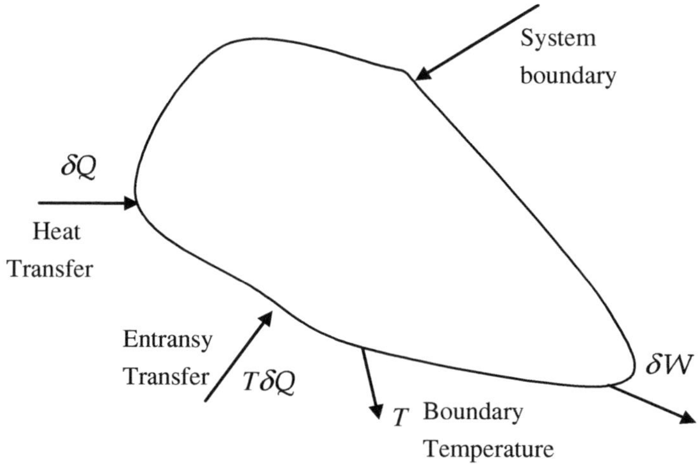

Fig. 22 Closed system

$$\int_1^2 \delta Q - \int_1^2 \delta W = E_2 - E_1 \tag{86}$$

where E is the energy, the subscripts 1 and 2 refer to the equilibrium states 1 and 2, respectively. Equation (86) can be written in the following form:

$$Q_{1,2} - W_{1,2} = E_2 - E_1 \tag{87}$$

where $Q_{1,2}$ and $W_{1,2}$ are respectively the heat transfer rate and work done on the environment when the state of the system changes from the state 1 to state 2. From Eq. (72) the second law of thermodynamics in terms of entransy for the system shown in Fig. 22 reads

$$\int_1^2 T\delta Q \geq G_2 - G_1 \tag{88}$$

The entransy transfer interaction between the closed system and its environment $T\delta Q$ is realized via heat transfer δQ at a boundary point of temperature T. According to the definition of the entransy generation expressed by Eq. (72), the entransy generation for the process that the system changes from state 1 to state 2 is written as follows

$$G_g = G_2 - G_1 - \int_1^2 T\delta Q \leq 0 \tag{89}$$

Or for the entransy dissipation, we have

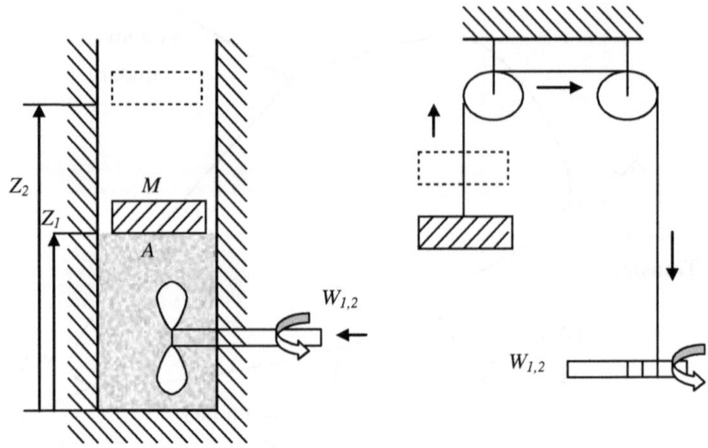

Fig. 23 Two apparatuses for lifting a weight to a pre-determined level

$$G_{diss} = \int_1^2 T\delta Q - (G_2 - G_1) \geq 0 \tag{90}$$

In Eqs. (89) and (90), the equality sign refers to the reversible process which does not generate entransy and the inequality sign corresponds to the irreversible process. It makes sense to describe process A as being more irreversible than process B whenever

$$(G_{diss})_A > (G_{diss})_B \tag{91}$$

In order to illustrate the relationship between the entransy dissipation and the destruction of available work, we consider a thermodynamic problem which is sketched in Fig. 23 which has been discussed by Bejan [13]. The apparatus consists of a piston-cylinder adiabatic enclosure filled with the ideal gas of mass m and initial conditions P_1, V_1 and T_1 (the pressure, volume and temperature of the gas at the initial position Z_1). The cylinder moves vertically, and the piston weight Mg (M is the mass of the piston, g is gravity acceleration) is always balanced by the gas pressure P_1A exerted on the piston face. Assume that the weight is lifted from an initial position $Z_1 = V_1/A$ (V_1 is the volume of the ideal gas at the position Z_1) to a higher position Z_2 by the movement of a paddlewheel inside the gas. After the piston rises to Z_2, the paddlewheel stops and the gas reaches internal equilibrium described by T_2, $V_2 = AZ_2$ and $P_2 = P_1 = Mg/A$.

Now we try to evaluate the work transfer $W_{1,2}$ required for lifting the weight by the left and right apparatus in Fig. 23.

The application of the first law of thermodynamics on the closed thermodynamic system containing the ideal gas and the weight, yields

$$Q_{1,2} + W_{1,2} = E_2 - E_1 \tag{92}$$

Note that $Q_{1,2} = 0$ and

$$\begin{aligned} E_2 - E_1 &= (U_1 - U_2)_{gas} + (E_2 - E_1)_{weight} \\ &= mc_v(T_2 - T_1) + Mg(Z_2 - Z_1) \end{aligned} \tag{93}$$

where U is the internal energy of the gas, m is the mass of the gas, c_v is the specific heat at constant volume. In deriving Eq. (93), the energy associated with lifting the gas and warming the piston has been neglected. Then using the state equation of the ideal gas and combing Eqs. (92) and (93) yield

$$W_{1,2} = \left(\frac{c_v}{R} + 1\right) Mg(Z_2 - Z_1) \tag{94}$$

where R is the ideal gas constant. Note that the work required for lifting the weight is proportional to the height difference. From Eq. (90) and concerning that $\delta Q = 0$, $P_1 = P_2$, the change of the entransy is expressed as

$$\Delta G = \int_1^2 mTdh = \frac{1}{2} mc_p (T_2^2 - T_1^2) \tag{95}$$

where c_p is the specific heat at constant pressure. The application of the state equation of ideal gas on Eq. (95) yields

$$\Delta G = \frac{1}{2} \frac{c_p M^2 g^2}{mR^2} (Z_2 - Z_1)(Z_2 + Z_1) \tag{96}$$

If $Z_2 - Z_1 \ll 2Z_1$, Eq. (96) becomes

$$\Delta G \approx \frac{c_p M^2 g^2}{mR^2} Z_1 (Z_2 - Z_1) \tag{97}$$

The important conclusion is that the lifting process is irreversible, and the entransy dissipation is proportional to the height difference $(Z_2 - Z_1)$.

Now we consider the process by which the weight is lifted, not by the expanding gas, but by some frictionless mechanical equipment powered electrically as shown in Fig. 23. Assume the environment temperature is T_1. For this case, the application of the first and second laws gives the following expression of the work required for lifting the weight:

$$W'_{1,2} = Mg(Z_2 - Z_1) \tag{98}$$

$$\Delta G' = T_1 Mg(Z_2 - Z_1) \tag{99}$$

From Eqs. (94) and (98), it is evident that the second lifting procedure requires less work than the first. Therefore, we can say that the reversible process is more efficient for delivering work. The additional work wasted by using the first lifting process instead of the second is

$$W_{1,2} - W'_{1,2} = \frac{c_v}{R} Mg(Z_2 - Z_1) \tag{100}$$

By the definition of the entransy generation given by Eq. (72), we have

$$G_g = \Delta G' - \Delta G = -\frac{M^2 g^2}{mR} Z_1 (Z_2 - Z_1) \tag{101}$$

Or the entransy dissipation is expressed as

$$G_{diss} = -G_g = \frac{M^2 g^2}{mR} Z_1 (Z_2 - Z_1) \tag{102}$$

From Eq. (102), Eq. (100) is rewritten as follows

$$W_{1,2} - W'_{1,2} = \left(\frac{c_v}{RT_1}\right) G_{diss} \tag{103}$$

Note that the wasted work for the irreversible process is proportional to the entransy dissipation. Therefore, the entransy dissipation can serve as a measure of the useful work destroyed in the irreversible process.

5.6 Open Systems

In this subsection, we consider an open system as shown in Fig. 24. In most cases the control surface is rigid. Therefore, the open system usually can not perform work by means of forces normal to the boundary. The only possibility for work transfer is by shear forces, such as a rotating shaft that penetrates the control surface. Note that the open system involves the mass, heat and entransy transfers. For mass transfer, we have

$$\dot{m}_{in} - \dot{m}_{out} = \frac{\partial M}{\partial t} \tag{104}$$

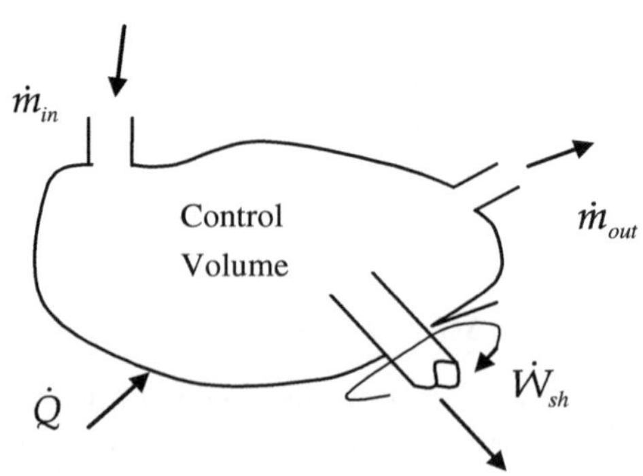

Fig. 24 Open thermodynamic system

where $\dot{m}$ is the mass flow rate, the subscripts *in* and *out* respectively represent the input and output, M is the total mass of the system, t is the time. For energy transfer, the first law of thermodynamics reads:

$$\dot{m}_{in}(h + \frac{1}{2}V^2 + gZ) - \dot{m}_{out}(h + \frac{1}{2}V^2 + gZ) + \dot{Q} - \dot{W}_{sh} = \frac{\partial E}{\partial t} \tag{105}$$

where h is the specific enthalpy, V the velocity, Z the height, E the system instantaneous energy integrated over the entire control volume.

The second law of the thermodynamics in terms of entransy and entransy dissipation is expressed as follows:

$$\dot{G}_g = \frac{\partial G}{\partial t} + T\dot{Q} + \dot{m}_{out}g_0 - \dot{m}_{in}g_0 \leq 0 \tag{106}$$

$$\dot{G}_{diss} = -\frac{\partial G}{\partial t} - T\dot{Q} - \dot{m}_{out}g_0 + \dot{m}_{in}g_0 \geq 0 \tag{107}$$

where g_0 is the specific entransy, $\dot{G}_g$ and $\dot{G}_{diss}$ are respectively the entransy generation and entransy dissipation rates. The entransy dissipation rate can be used for quantifying a system's irreversibility. If systems A and B operate so that $\dot{G}_{diss,A} > \dot{G}_{diss,B}$, we can say that system A operates more irreversibly than system B.

Now we consider the steady operation of a steam turbine as shown in Fig. 25. The steady operation implies that the system properties such as density, specific energy, and specific entransy are independent on the time. Across the turbine, the steam pressure drops from a fixed inlet pressure P_{in} to a fixed outlet pressure P_{out}. Therefore, concerning the steady operation condition, Eq. (104) reduces to

$$\dot{m}_{in} = \dot{m}_{out} = \dot{m} \tag{108}$$

Assume the system under consideration is adiabatic, the first law and the second law are expressed as

$$\dot{W}_{sh} = \dot{m}(h_{in} - h_{out}) \tag{109}$$

$$\Delta\dot{G} = \dot{m}(g_{0,out} - g_{0,in}) \geq 0 \tag{110}$$

where $\Delta\dot{G}$ is the entransy change. In writing Eq. (109), the kinetic and gravitational energy changes related to the inlet and outlet streams have been neglected.

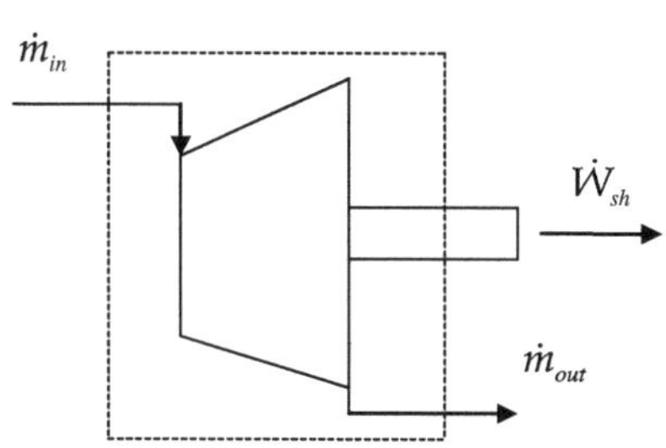

Fig. 25 A turbine with steady flow

Assume the working fluid is the ideal gas. Denote the inlet and outlet temperatures of the gas are T_{in} and T_{out}, respectively. Then we have

$$\Delta\dot{G} = \dot{m}(T_{out}h_{out} - T_{in}h_{in}) \tag{111}$$

Now we consider the limiting situation that the flow of steam through the turbine is in the reversible mode. When P_{in} and P_{out} are fixed, a reversible turbine delivers the maximum possible shaft power per mass flow rate. In the reversible case, Eqs. (109) and (111) reduce to

$$\dot{W}_{sh,\max} = \dot{m}(h_{in} - h_{out,rev}) \tag{112}$$

$$\Delta\dot{G}' = \dot{m}(T_{out,rev}h_{out,rev} - T_{in}h_{in}) \tag{113}$$

Combining Eqs. (109) and (112) gives

$$\dot{W}_{sh,\max} - \dot{W}_{sh} = \dot{m}(h_{out} - h_{out,rev}) = \dot{m}c_p(T_{out} - T_{out,rev}) \tag{114}$$

From Eqs. (111) and (114), the entransy dissipation is written as

$$\dot{G}_{diss} = \dot{m}(T_{out}h_{out} - T_{out,rev}h_{out,rev}) = \dot{m}c_p(T_{out}^2 - T_{out,rev}^2) \tag{115}$$

From Eqs. (114) and (115), we obtain

$$\dot{W}_{sh,\max} - \dot{W}_{sh} = \frac{1}{T_{out} + T_{out,rev}}\dot{G}_{diss} \tag{116}$$

Therefore, the shaft work consumed by the irreversibility in the turbine is proportional to the entransy dissipation.

Note that the concept of entropy can be replaced with entransy for describing the second law of thermodynamics. Furthermore, we show that the useful work destroyed by the irreversible process in the closed and open systems is proportional to the entransy dissipation. In comparison with entropy, the entransy has a definite physical meaning and the principle of minimum entransy dissipation rate avoids the contradiction between the principle of minimum entropy production and the classical Fourier heat conduction theory. Therefore, we anticipate that the concepts of entransy and entransy dissipation may help us to gain more profound insight on thermodynamics in particular and on science in general.

6 The Application of Entransy Dissipation Theory in Heat Convection

In this section, the entransy dissipation theory developed in Sect. 4 is applied to derive the expression of the entransy dissipation of heat convection.

6.1 Introduction

The entropy and entropy generation can help us to deeply understand the momentum and heat transfer [12, 83]. Bejan realized that in order to improve the performance of the heat transfer enhancement or thermal insulation equipments, one need to reduce the entropy generation rate [12]. Similarly, according to the definition of entransy dissipation given in Sect. 4.4, it is required to minimize the entransy dissipation rate for achieving the best heat transfer enhancement and thermal insulation. Therefore, it is of great value to derive the expression of the local entransy dissipation rate for heat convection. Xu et al. have managed to get an expression of the local entransy dissipation rate for heat convection [5]. However, it lacks the theoretical basis in this derivation. In Sect. 5.2 with the help of the second law of thermodynamics in terms of entransy and entransy dissipation established in Sect. 4.3 we will make the derivation more rigorously.

6.2 Local Thermodynamic Entransy Dissipation in Heat Convection

The infinitesimal element as shown in Fig. 26 is an open thermodynamic system, where $[v_x, v_y]^T$ is the velocity, $[q_x, q_y]^T$ is the heat flux. For this system, we assume that the thermodynamic state is irrelevant with the position, but relevant with time. By the second law of thermodynamics for the open system expressed as Eq. (104) the local thermodynamic entransy dissipation rate $\dot{G}_g$ per unit volume in the infinitesimal element is expressed as

$$\begin{aligned}
\dot{G}_g dxdy = &-\left(T - T - \frac{\partial T}{\partial x}dx\right)q_x dy - \left(T - T - \frac{\partial T}{\partial y}dy\right)q_y dx \\
&+ T\left(q_x - q_x - \frac{\partial q_x}{\partial x}dx\right)dy + T\left(q_y - q_y - \frac{\partial q_y}{\partial y}dy\right)dx \\
&+ \left(g_0 + \frac{\partial g_0}{\partial x}dx\right)\left(v_x + \frac{\partial v_x}{\partial x}dx\right)\left(\rho + \frac{\partial \rho}{\partial x}dx\right)dy \\
&+ \left(g_0 + \frac{\partial g_0}{\partial y}dy\right)\left(v_y + \frac{\partial v_y}{\partial y}dy\right)\left(\rho + \frac{\partial \rho}{\partial y}dy\right)dx \\
&- \rho v_x g_0 dy - \rho v_y g_0 dx + \frac{\partial(\rho g_0)}{\partial t}dxdy
\end{aligned} \tag{117}$$

where g_0 is the specific entransy, ρ is the density of fluid. The first four terms in the right side of Eq. (117) account for the entransy exchanged with the environment through the element boundary, the following four terms represent the entransy transported by convection, the last term stands for the variation of the entransy with respect to time. Rearranging Eq. (117) yields

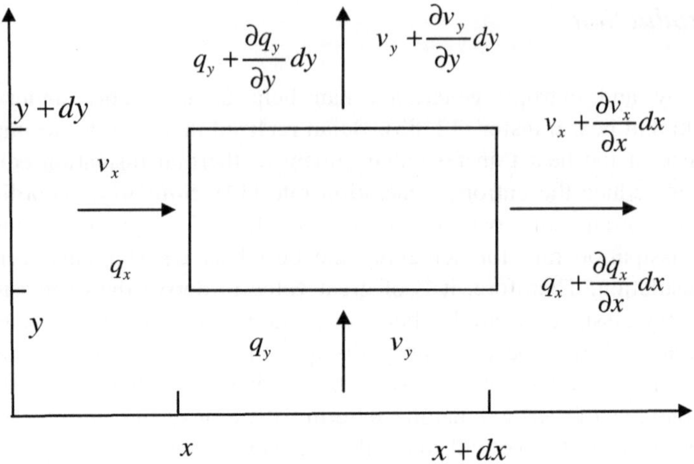

Fig. 26 Two-dimensional infinitesimal element for heat convection

$$\begin{aligned} \dot{G}_g = & \left(q_x \frac{\partial T}{\partial x} + q_y \frac{\partial T}{\partial y} \right) - T \left(\frac{\partial q_x}{\partial x} + \frac{\partial q_y}{\partial y} \right) + \rho \left(\frac{\partial g_0}{\partial t} + v_x \frac{\partial g_0}{\partial x} + v_y \frac{\partial g_0}{\partial y} \right) \\ & + g_0 \left[\frac{\partial \rho}{\partial t} + \rho \left(\frac{\partial v_x}{\partial x} + \frac{\partial v_y}{\partial y} \right) + v_x \frac{\partial \rho}{\partial x} + v_y \frac{\partial \rho}{\partial y} \right] \end{aligned} \tag{118}$$

Making use of the following mass balance equation,

$$\frac{\partial \rho}{\partial t} + v_x \frac{\partial \rho}{\partial x} + v_y \frac{\partial \rho}{\partial y} + \rho \left(\frac{\partial v_x}{\partial x} + \frac{\partial v_y}{\partial y} \right) = 0 \tag{119}$$

where ρ is the density of fluid, Eq. (118) is simplified to

$$\dot{G}_g = \boldsymbol{q} \cdot \nabla T - T \nabla \cdot \boldsymbol{q} + \rho \frac{Dg_0}{Dt} \tag{120}$$

where $D/DT = \partial/\partial t + v_x \partial/\partial x + v_y \partial/\partial y$ is the material derivative. By the following canonical relation of thermodynamics,

$$\rho \frac{Dg_0}{Dt} = -T\rho \frac{Du}{Dt} + P \frac{T}{\rho} \frac{D\rho}{Dt} \tag{121}$$

where u is the specific internal energy, P is the pressure, as well as the definition of material derivative. The energy balance equation for heat convection reads,

$$\rho \frac{Du}{Dt} = -\nabla \cdot \boldsymbol{q} - P(\nabla \cdot \boldsymbol{v}) + \mu \Phi \tag{122}$$

where μ is the viscosity, Φ is the dissipation function and can be expressed as

$$\Phi = 2\left[\left(\frac{\partial v_x}{\partial x}\right)^2 + \left(\frac{\partial v_y}{\partial y}\right)^2\right] + \left(\frac{\partial v_x}{\partial y} + \frac{\partial v_y}{\partial x}\right)^2 \tag{123}$$

Combining Eqs. (121) and (122) and using Eq. (119), we obtain

$$\rho\frac{Dg_0}{Dt} = T\nabla \cdot \boldsymbol{q} - T\mu\Phi \tag{124}$$

Substituting Eq. (124) into Eq. (120) yields

$$\dot{G}_g = \boldsymbol{q} \cdot \nabla T - \mu T\Phi \tag{125}$$

The application of the Fourier law, $\mathbf{q} = -k \cdot \nabla T$, on Eq. (125) gives,

$$\dot{G}_g = -k(\nabla T)^2 - \mu T\Phi \tag{126}$$

where k is the thermal conductivity. Since $\dot{G}_{diss} = -\dot{G}_g$, we have

$$\dot{G}_{diss} = k(\nabla T)^2 + \mu T\Phi \tag{127}$$

Thus we obtained the formula for calculating the local entransy dissipation rate in heat convection. It is evident that the expression of the local entransy dissipation rate in the three-dimensional case shares the same form as Eq. (127).

6.3 Entransy Dissipation for Internal Flow

From Eq. (127), one can see that if the temperature and velocity fields in heat convection problems are known, the local entransy dissipation rate can be obtained exactly. Unfortunately, it is seldom to get the exact solution of heat convection problems. In many cases only the heat transfer correlations are available. Therefore it is of importance to develop expressions of the entransy dissipation rate by the knowledge of heat transfer correlations for heat convection. In this section we attempt to establish such a formula for internal flow in a duct with arbitrary cross section. Consider a flow passage enclosed by two cross sections with distance dx. The temperatures of the duct wall and fluid inside the duct are $T + \Delta T$ and T, respectively. The heat transfer rate from the duct wall to the fluid per unit length is $\dot{q}$. The fluid inside the duct flows with friction in x-direction, thus the pressure gradient $dP/dx \leq 0$. In this heat transfer process, assume there are two irreversibilities. One is the heat conduction from the duct wall to the fluid. The other irreversibility is flow friction. As discussed in Sect. 4.4 the entransy dissipation rate per unit length $\left(\partial \dot{G}_{diss}/\partial x\right)_T$ induced by the heat conduction is

$$\left(\frac{\partial \dot{G}_{diss}}{\partial x}\right)_T = \dot{q}\Delta T \tag{128}$$

Assume the fluid is incompressible. As discussed in Sect. 4.2, the corresponding entransy dissipation rate per unit length $\left(\partial \dot{G}_{diss}/\partial x\right)_P$ is expressed as

$$\left(\frac{\partial \dot{G}_{diss}}{\partial x}\right)_P = -\frac{\dot{m}T}{\rho}\left(\frac{dP}{dx}\right) \tag{129}$$

where ρ is the density of the fluid, $\dot{m}$ is the mass flow rate. Thus the total entransy dissipation rate per unit length $\left(\partial \dot{G}_{diss}/\partial x\right)$ is

$$\left(\frac{\partial \dot{G}_{diss}}{\partial x}\right) = \dot{q}\Delta T - \frac{\dot{m}T}{\rho}\left(\frac{dP}{dx}\right) \tag{130}$$

The Stanton number correlation reads

$$S_t = \frac{\dot{q}}{p\Delta T c_p G} \tag{131}$$

where $G = \dot{m}/A$, A is the cross sectional area of the duct, c_p is the specific heat at constant pressure, p is the wetted perimeter. And the friction factor correlation is

$$f = -\frac{\rho D}{2G^2}\frac{dP}{dx} \tag{132}$$

where D is the hydraulic diameter. By Eqs. (131) and (132), Eq. (130) arrives at

$$\frac{\partial \dot{G}_{diss}}{\partial x} = \frac{q^2}{4\dot{m}c_p}\frac{D}{S_t} + \frac{2\dot{m}^3 T}{\rho^2}\frac{f}{DA^2} \tag{133}$$

Thus we obtain the formula for calculating the derivative of the entransy dissipation rate with respect to the position variable for the forced convection in duct flow. Note that the higher Stanton number means the smaller entransy dissipation rate, while a high friction factor leads to the increase of the entransy dissipation rate.

6.4 Entransy Dissipation for External Flow

Consider an external steady flow that a solid body with arbitrary shape is suspended in a uniform stream with velocity U_∞ and absolute temperature T_∞. For the convenience of analysis, the stream tube with cylindrical shape and large enough volume around the solid body as shown in Fig. 27 is taken as the thermodynamic system. The surface temperature of the solid body is T_w. In the regions near the stream tube surface and the external part, the influence of the solid body on the flow is neglected. There is no heat transfer between the solid surface and its external environment. The irreversible dissipation is only caused by the heat

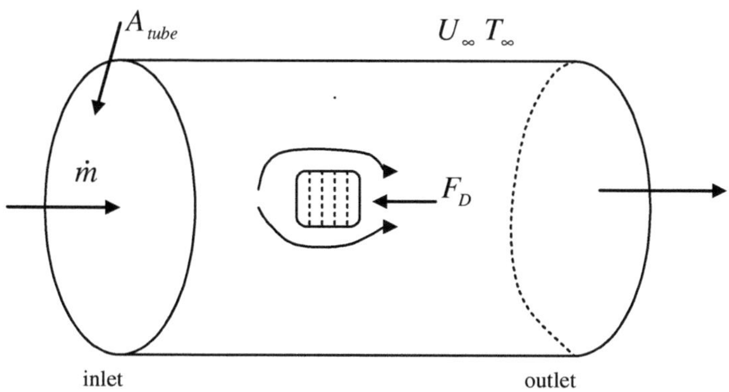

Fig. 27 Schematic diagram for external flow

conduction and flow friction. For the considered open thermodynamic system, the mass and energy conservations are expressed as,

$$\dot{m}_{in} = \dot{m}_{out} = \dot{m} \tag{134}$$

$$\dot{m}h_{in} + \iint_A \dot{q}ds - \dot{m}h_{out} = 0 \tag{135}$$

where the subscripts "in" and "out" respectively indicate the inlet and outlet, $\dot{q}$ is the local heat transfer rate between the solid body and the stream, A is the surface area of the solid body, $\dot{m}$is the mass flow rate. For the heat conduction across the nonzero temperature difference $T_w - T_\infty$, the entransy dissipation rate $\dot{G}_{diss,T}$ is written as

$$\dot{G}_{diss,T} = \iint_A \dot{q}(T_w - T_\infty)dA \tag{136}$$

where A is the surface area of the solid body. For the flow friction, the entransy dissipation rate $\dot{G}_{diss,P}$ is

$$\dot{G}_{diss,P} = -\frac{\dot{m}T_\infty}{\rho_\infty}(P_{out} - P_{in}) \tag{137}$$

where ρ_∞ is the density of the fluid, P is the pressure. Here we assume the fluid is incompressible. Therefore, the total entransy dissipation rate is written as

$$\dot{G}_{diss} = \iint_A \dot{q}(T_w - T_\infty)dA - \frac{\dot{m}T_\infty}{\rho_\infty}(P_{out} - P_{in}) \tag{138}$$

Since

$$\dot{m} = A_{tube}\rho_\infty U_\infty \tag{139}$$

and from the force balance on the control volume, we have

$$F_D = A_{tube}(P_{in} - P_{out}) \tag{140}$$

where F_D is the drag force, A_{tube} is the cross sectional area of the cylindrical tube. Substituting Eq. (140) into Eq. (138) yields

$$\dot{G}_{diss} = \iint_A \dot{q}(T_w - T_\infty)dA + T_\infty F_D U_\infty \tag{141}$$

If the body temperature is uniform, then the temperature difference $T_w - T_\infty$ is constant, thus Eq. (141) becomes

$$\dot{G}_{diss} = (T_w - T_\infty)\iint_A \dot{q}dA + T_\infty F_D U_\infty \tag{142}$$

Since

$$\iint_A \dot{q}dA = Q = \bar{h}A(T_w - T_\infty) \tag{143}$$

where $\bar{h}$ is the average heat transfer coefficient based on A, Eq. (141) arrives at

$$\dot{G}_{diss} = \bar{h}A(T_w - T_\infty)^2 + T_\infty F_D U_\infty \tag{144}$$

Note that when the body-ambient temperature difference is fixed, the only way to minimize the heat transfer contribution to the entransy dissipation rate is by decreasing the thermal conductance $\bar{h}A$.

The other extreme is that the heat transfer rate $\dot{q}$ is uniform around the solid body. In this case Eq. (141) becomes

$$\dot{G}_{diss} = \dot{q}^2\iint_A \frac{1}{h}dA + T_\infty F_D U_\infty \tag{145}$$

where h is the local heat transfer coefficient. Note that in order to reduce the heat transfer entransy dissipation rate, we must increase the heat transfer coefficient. In other words, the enhancement of the heat transfer from the solid body to the external fluid can decrease the entransy dissipation rate.

6.5 Variational Principles in Terms of Entransy

In this section, our work presented in [93] is reviewed. In this work a variational principle for heat conduction is firstly formulated which results in the steady state heat conduction equation established from the Fourier law. Furthermore based on the thermodynamics in terms of entransy a more general functional is defined for incompressible fluids. We show that extremizing this functional gives rise to the state that the substantive derivatives of the temperature and velocity field vanish in Navier–Stokes–Fourier equations. In this sense one may conclude that this variational principle is consistent with the Navier–Stokes–Fourier equations. Therefore the variational principle developed in the present work demonstrates a great advantage over the minimum entropy production principle.

6.5.1 Introduction

Variational principles have played an important role in the development of physical sciences, such as the Fermat's principle in optics and the principle of least action in mechanics. In thermodynamics, much effort has been devoted to establishing the variational principles for nonequilibrium systems, which has led to several variational principles, such as the principle of minimum rate of entropy production due to Prigogine, the principle of maximum rate of entropy production, two principles of minimum dissipation [59, 94–97]. The principle of minimum rate of entropy production which reads "A steady state has the minimum rate of entropy production with respect to other possible states with the same boundary conditions" has inspired the developments of the finite time thermodynamics [79–81], the entropy generation minimization approach which has been widely applied in optimization designs of thermal systems [12, 13] and the constructal theory [98]. However, it has also caused lots of controversies. It was found that for the reacting systems displaced from equilibrium by a matter flux across the boundaries the state of minimum entropy production differs from the steady state, even in the near equilibrium regime [99]. Tykodi showed that the steady states of a family of chemical reaction schemes, heat conduction, diffusion and Poiseuille flow of a fluid fail to satisfy the principle of minimum entropy production [100]. Hillert and Ågren indicated that Prigogine's principle of minimum entropy production is not strictly valid except for a very special case [101]. Barragan found that the steady state described by Newton's cooling law can not lead to the minimum entropy production [102]. Attard pointed that the principle of minimum entropy production is ill conceived and is not directly relevant to nonequilibrium systems [103]. Recently, Bertola and Cafaro found that the minimum entropy production principle produces the field equations that do not agree with the balance equations of continuum mechanics for the steady heat conduction and shear flow of an incompressible fluid [84].

In fact not only the minimum entropy production principle but also the second law of thermodynamics exhibits the inconsistency with the fluid mechanics [104]. During the applications of the concepts of entropy and entropy generation in thermal engineering, some paradoxes and inconsistence have also been found. Herwig showed that for heat transfer in rough pipes there exists a big difference between the optimization results obtained from the entropy generation minimization and the performance evaluation criterion based on the heat transfer theory [83]. By analyzing the heat exchanger effectiveness and entropy generation of eighteen types of heat exchangers Shah and Skiepko found that the heat exchanger effectiveness can be maximum, minimum or in between when the entropy generation achieves its minimum value [85]. We also found that the dimensionless entropy generation defined by Bejan [57] is not compatible with the exchanger effectiveness which describes the heat exchanger performance [8].

With the aim to resolve the 'entropy generation paradox' induced by the entropy generation number Guo et al. defined two new physical concepts called entransy and entransy dissipation for describing the heat transfer ability and irreversibility of heat transfer, respectively [86]. Based on the concept of entransy dissipation, an equivalent thermal resistance of heat exchanger was defined and the relationship between the exchanger effectiveness and the thermal resistance was established [87]. By introducing a new approach to non-dimensionalising the entransy dissipation in heat exchanger we defined an entransy dissipation number which can serve as the heat exchanger performance evaluation criterion [8]. Liu et al. developed a minimum entransy dissipation principle for the optimization of transport networks [88]. Chen et al. applied the entransy dissipation extremum principle to optimize the convective heat transfer processes and made a comparison with the minimum entropy generation principle [89]. In Ref. [90], the entransy dissipation theory was extended to deal with the coupled heat and mass transfer processes in evaporative cooling systems. In Ref. [91], Cheng and Liang defined the entransy flux and entransy function for the thermal radiation in enclosures with opaque surfaces, and developed the minimum principle of radiative entransy. Chen et al. proposed an entransy dissipation rate minimization approach for the disc cooling system and investigated the influence of various system parameters on the entransy dissipation rate [92]. Recently, we defined the entransy and entransy dissipation from the thermodynamic point of view and showed that the entransy is a state variable and can be employed to describe the second law of thermodynamics [4].

6.5.2 The Net Rate of Entransy Flowed to Fluid Particle

We consider a small element of fluid with sides δx, δy and δz. The centre of the element is located at position (x, y, z). The element under consideration is so small that the temperature field and heat flux components at the faces can be expressed accurately enough by means of the first two terms of a Taylor series expansion as shown in Fig. 28.

The net rate of entransy flowed into the fluid particle due to heat flow in the x-direction is given as follows:

$$\left[\left(q_x - \frac{\partial q_x}{\partial x}\frac{1}{2}\delta x\right)\left(T - \frac{\partial T}{\partial x}\frac{1}{2}\delta x\right) - \left(q_x + \frac{\partial q_x}{\partial x}\frac{1}{2}\delta x\right)\left(T + \frac{\partial T}{\partial x}\frac{1}{2}\delta x\right)\right]\delta y \delta x \quad (146)$$

where q_x is the x-component of the heat flux vector $(q_x, q_y, q_z)^T$, T is the temperature. Neglecting the higher order terms in Eq. (146), we have

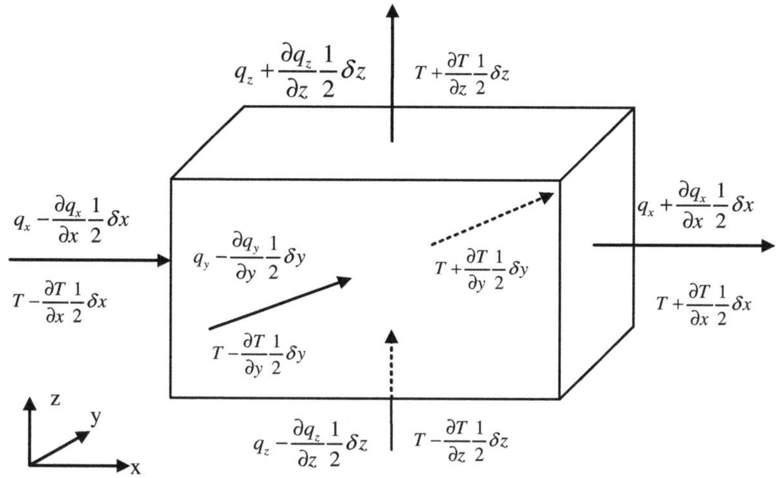

Fig. 28 Heat flux and temperature on surfaces of a fluid element

$$\left(-q_x\frac{\partial T}{\partial x}-\frac{\partial q_x}{\partial x}T\right)\delta x\delta y\delta z \tag{147}$$

Similarly, the net rates of entransy flowed into the fluid particle due to heat flows in the $y-$ and $z-$ direction are expressed as

$$\left(-q_y\frac{\partial T}{\partial y}-\frac{\partial q_y}{\partial y}T\right)\delta x\delta y\delta z \text{ and } \left(-q_z\frac{\partial T}{\partial z}-\frac{\partial q_z}{\partial z}T\right)\delta x\delta y\delta z \tag{148}$$

Thus the total rate of entransy F_T added to the fluid particle per unit volume due to heat flow across its boundaries is

$$F_T=-q_x\frac{\partial T}{\partial x}-\frac{\partial q_x}{\partial x}T-q_y\frac{\partial T}{\partial y}-\frac{\partial q_y}{\partial y}T-q_z\frac{\partial T}{\partial z}-\frac{\partial q_z}{\partial z}T \tag{149}$$

The application of the Fourier law on Eq. (149) yields

$$F_T=k\left[\left(\frac{\partial T}{\partial x}\right)^2+\left(\frac{\partial T}{\partial y}\right)^2+\left(\frac{\partial T}{\partial z}\right)^2\right]+Tk\left[\frac{\partial^2 T}{\partial x^2}+\frac{\partial^2 T}{\partial y^2}+\frac{\partial^2 T}{\partial z^2}\right] \tag{150}$$

where k is the thermal conductivity and is assumed to be a constant.

6.5.3 The Consumed Entransy Rate Due to the Work Done on Fluid Particle

The x-components of the forces due to pressure p and stress components τ_{xx}, τ_{yx} and τ_{zx} are shown in Fig. 29. According to the definition of the entransy

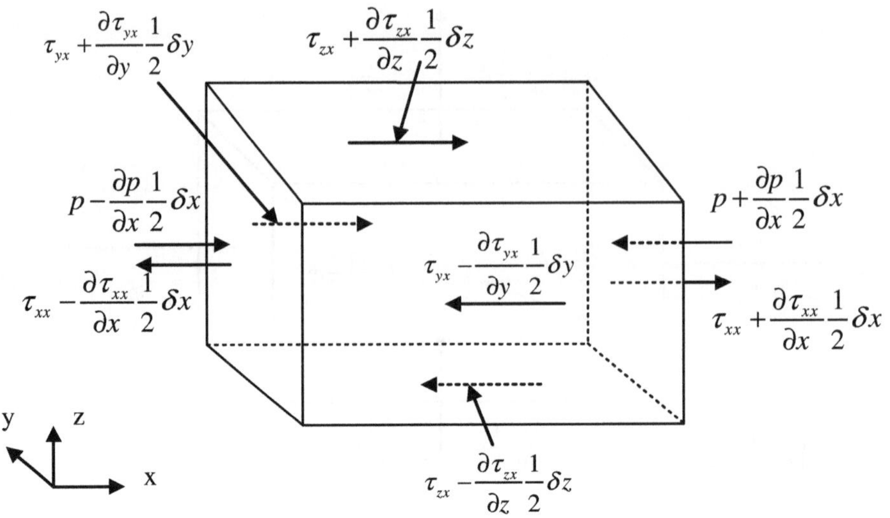

Fig. 29 Stress components in the x-direction

contributing to deliver useful work in the Carnot's cycle, the consumed entransy due to work done on the fluid particle by the forces in the x-direction is given by

$$\left[\left(pv_x-\frac{\partial(pv_x)}{\partial x}\frac{1}{2}\delta x\right)-\left(\tau_{xx}v_x-\frac{\partial(\tau_{xx}v_x)}{\partial x}\frac{1}{2}\delta x\right)-\left(pv_x+\frac{\partial(pv_x)}{\partial x}\frac{1}{2}\delta x\right)\right.$$
$$\left.+\left(\tau_{xx}v_x+\frac{\partial(\tau_{xx}v_x)}{\partial x}\frac{1}{2}\delta x\right)\right]T\delta y\delta z+\left[-\left(\tau_{yx}v_x-\frac{\partial(\tau_{yx}v_x)}{\partial y}\frac{1}{2}\delta y\right)+\left(\tau_{yx}v_x+\frac{\partial(\tau_{yx}v_x)}{\partial y}\frac{1}{2}\delta y\right)\right]T\delta x\delta z$$
$$+\left[-\left(\tau_{zx}v_x-\frac{\partial(\tau_{zx}v_x)}{\partial z}\frac{1}{2}\delta z\right)+\left(\tau_{zx}v_x+\frac{\partial(\tau_{zx}v_x)}{\partial z}\frac{1}{2}\delta z\right)\right]T\delta x\delta y$$

where v_x, v_y and v_z are the x-, y- and z-components of the velocity vector, respectively. Rearranging the above equations gives the following consumed entransy due to the work done on the fluid particle in the x-direction

$$\left[\frac{\partial(v_x(-p+\tau_{xx}))}{\partial x}+\frac{\partial(v_x\tau_{yx})}{\partial y}+\frac{\partial(v_x\tau_{zx})}{\partial z}\right]T\delta x\delta y\delta z \tag{151}$$

Similarly, the entransy due to the work done on the fluid particle in the y- and z-direction is expressed as

$$\left[\frac{\partial(v_y\tau_{xy})}{\partial x}+\frac{\partial(v_y(-p+\tau_{yy}))}{\partial y}+\frac{\partial(v_y\tau_{zy})}{\partial z}\right]T\delta x\delta y\delta z \tag{152}$$

$$\left[\frac{\partial(v_z\tau_{xz})}{\partial x}+\frac{\partial(v_z\tau_{yz})}{\partial y}+\frac{\partial(v_z(-p+\tau_{zz}))}{\partial z}\right]T\delta x\delta y\delta z \tag{153}$$

Thus the total consumed entransy F_W due to the work done per unit volume on the fluid particle can be written as

$$\begin{aligned} F_W = & \left[\frac{\partial(v_x(-p+\tau_{xx}))}{\partial x} + \frac{\partial(v_x\tau_{yx})}{\partial y} + \frac{\partial(v_x\tau_{zx})}{\partial z} \right. \\ & + \frac{\partial(v_y\tau_{xy})}{\partial x} + \frac{\partial(v_y(-p+\tau_{yy}))}{\partial y} + \frac{\partial(v_y\tau_{zy})}{\partial z} \\ & \left. + \frac{\partial(v_z\tau_{xz})}{\partial x} + \frac{\partial(v_z\tau_{yz})}{\partial y} + \frac{\partial(v_z(-p+\tau_{zz}))}{\partial z} \right] T \end{aligned} \tag{154}$$

6.5.4 The Extremum Principle in Terms of Entransy for Heat Conduction

From Eqs. (150) we define the following functional for heat conductions:

$$\int_V F_T dV \tag{155}$$

Then extremizing the functional (155) is equivalent to the following Euler–Lagrange equation:

$$\frac{\partial}{\partial x}\left(\frac{\partial F_T}{\partial(\frac{\partial T}{\partial x})}\right) + \frac{\partial}{\partial y}\left(\frac{\partial F_T}{\partial(\frac{\partial T}{\partial y})}\right) + \frac{\partial}{\partial z}\left(\frac{\partial F_T}{\partial(\frac{\partial T}{\partial z})}\right) - \frac{\partial F_T}{\partial T} = 0 \tag{156}$$

Substituting Eq. (150) into Eq. (156) yields

$$\frac{\partial^2 T}{\partial x^2} + \frac{\partial^2 T}{\partial y^2} + \frac{\partial^2 T}{\partial z^2} = 0 \tag{157}$$

which is nothing more than the governing equation of the steady heat conduction established by the Fourier law. Therefore, the extremum of the functional (155) gives rise to the steady state of heat conductions. In this point, it demonstrates an obvious advantage over the minimum entropy production principle which can not lead to the Fourier steady heat conduction equation.

6.5.5 The Extremum Principle in Terms of Entransy for Heat Transfer and Fluid Flow

We restrict ourselves to the incompressible flow and assume the following constitutive relationship:

$$\tau_{xx} = 2\mu\frac{\partial v_x}{\partial x},\ \tau_{yy} = 2\mu\frac{\partial v_y}{\partial y},\ \tau_{zz} = 2\mu\frac{\partial v_z}{\partial z}$$

$$\tau_{xy} = \tau_{yx} = \mu\left(\frac{\partial v_x}{\partial y} + \frac{\partial v_y}{\partial x}\right),\ \tau_{xz} = \tau_{zx} = \mu\left(\frac{\partial v_x}{\partial z} + \frac{\partial v_z}{\partial x}\right),\ \tau_{yz} = \tau_{zy} = \mu\left(\frac{\partial v_y}{\partial z} + \frac{\partial v_z}{\partial y}\right)$$

where μ is the dynamic viscosity and assumed to be constant. By analogy with the description of the Carnot's cycle in terms of entransy we define the following functional:

$$\int_V F dV \tag{158}$$

with

$$\begin{aligned} F = {} & F_T - F_W + \lambda_1 T\left(\tau_{xx} - 2\mu\frac{\partial v_x}{\partial x}\right) + \lambda_2 T\left(\tau_{xy} - \mu\left(\frac{\partial v_x}{\partial y} + \frac{\partial v_y}{\partial x}\right)\right) \\ & + \lambda_3 T\left(\tau_{xz} - \mu\left(\frac{\partial v_x}{\partial z} + \frac{\partial v_z}{\partial z}\right)\right) + \lambda_4 T\left(\tau_{yy} - 2\mu\frac{\partial v_y}{\partial y}\right) \\ & + \lambda_5 T\left(\tau_{zz} - 2\mu\frac{\partial v_z}{\partial z}\right) + \lambda_6 T\left(\tau_{yz} - \left(\frac{\partial v_y}{\partial z} + \frac{\partial v_z}{\partial y}\right)\right) \end{aligned} \tag{159}$$

where λ_i $(i = 1, 2, \cdots, 5)$ are the Lagrange multipliers. Extremizing the functional defined by Eq. (158) gives the following Euler–Lagrange equations:

$$\frac{\partial}{\partial x}\left(\frac{\partial F}{\partial(\frac{\partial v_x}{\partial x})}\right) + \frac{\partial}{\partial y}\left(\frac{\partial F}{\partial(\frac{\partial v_x}{\partial y})}\right) + \frac{\partial}{\partial z}\left(\frac{\partial F}{\partial(\frac{\partial v_x}{\partial z})}\right) - \frac{\partial F}{\partial v_x} = 0 \tag{160}$$

$$\frac{\partial}{\partial x}\left(\frac{\partial F}{\partial(\frac{\partial v_y}{\partial x})}\right) + \frac{\partial}{\partial y}\left(\frac{\partial F}{\partial(\frac{\partial v_y}{\partial y})}\right) + \frac{\partial}{\partial z}\left(\frac{\partial F}{\partial(\frac{\partial v_y}{\partial z})}\right) - \frac{\partial F}{\partial v_y} = 0 \tag{161}$$

$$\frac{\partial}{\partial x}\left(\frac{\partial F}{\partial(\frac{\partial v_z}{\partial x})}\right) + \frac{\partial}{\partial y}\left(\frac{\partial F}{\partial(\frac{\partial v_z}{\partial y})}\right) + \frac{\partial}{\partial z}\left(\frac{\partial F}{\partial(\frac{\partial v_z}{\partial z})}\right) - \frac{\partial F}{\partial v_z} = 0 \tag{162}$$

$$\frac{\partial}{\partial x}\left(\frac{\partial F}{\partial(\frac{\partial T}{\partial x})}\right) + \frac{\partial}{\partial y}\left(\frac{\partial F}{\partial(\frac{\partial T}{\partial y})}\right) + \frac{\partial}{\partial z}\left(\frac{\partial F}{\partial(\frac{\partial T}{\partial z})}\right) - \frac{\partial F}{\partial T} = 0 \tag{163}$$

Substituting Eq. (159) into Eq. (160) yields

$$
\begin{aligned}
&- T\frac{\partial p}{\partial x} + T\frac{\partial \tau_{xx}}{\partial x} + T\frac{\partial \tau_{zx}}{\partial z} + T\frac{\partial \tau_{xy}}{\partial y} - T\frac{\partial \tau_{xx}}{\partial x} - \tau_{xx}\frac{\partial T}{\partial x} \\
&- 2\mu T\frac{\partial \lambda_1}{\partial x} - 2\mu\lambda_1\frac{\partial T}{\partial x} - \frac{\partial \tau_{xy}}{\partial y}T - \frac{\partial T}{\partial y}\tau_{xy} - T\mu\frac{\partial \lambda_2}{\partial y} \\
&- \mu\lambda_2\frac{\partial T}{\partial y} - T\frac{\partial \tau_{zx}}{\partial z} - \tau_{zx}\frac{\partial T}{\partial z} - \mu T\frac{\partial \lambda_3}{\partial z} - \mu\lambda_3\frac{\partial T}{\partial z} = 0
\end{aligned} \tag{164}
$$

Let $\lambda_1 = -\tau_{xx}/(2\mu)$, $\lambda_2 = -\tau_{xy}/\mu, \lambda_3 = -\tau_{zx}/\mu$, Eq. (164) becomes

$$\frac{\partial \tau_{xx}}{\partial x} + \frac{\partial \tau_{yx}}{\partial y} + \frac{\partial \tau_{zx}}{\partial z} - \frac{\partial p}{\partial x} = 0 \tag{165}$$

Similarly substituting Eq. (159) into Eqs. (161–163) and setting $\lambda_4 = -\tau_{yy}/(2\mu), \lambda_5 = -\tau_{zz}/(2\mu), \lambda_6 = -\tau_{zy}/\mu$, yield

$$\frac{\partial \tau_{xy}}{\partial x} + \frac{\partial \tau_{yy}}{\partial y} + \frac{\partial \tau_{zy}}{\partial z} - \frac{\partial p}{\partial y} = 0 \tag{166}$$

$$\frac{\partial \tau_{xz}}{\partial x} + \frac{\partial \tau_{yz}}{\partial y} + \frac{\partial \tau_{zz}}{\partial z} - \frac{\partial p}{\partial z} = 0 \tag{167}$$

$$\frac{\partial}{\partial x}\left(k\frac{\partial T}{\partial x}\right) + \frac{\partial}{\partial y}\left(k\frac{\partial T}{\partial y}\right) + \frac{\partial}{\partial z}\left(k\frac{\partial T}{\partial z}\right) + \mu\Phi = 0 \tag{168}$$

where Φ is the dissipation function and is equal to:

$$\Phi = \mu\left\{2\left[\left(\frac{\partial v_x}{\partial x}\right)^2 + \left(\frac{\partial v_y}{\partial y}\right)^2 + \left(\frac{\partial v_z}{\partial z}\right)^2\right] + \left(\frac{\partial v_x}{\partial y} + \frac{\partial v_y}{\partial x}\right)^2 + \left(\frac{\partial v_x}{\partial z} + \frac{\partial v_z}{\partial x}\right)^2 + \left(\frac{\partial v_y}{\partial z} + \frac{\partial v_z}{\partial y}\right)^2\right\}$$

From Eqs. (165–168), one can see that extremizing the functional (158) gives rise to the Stokes equations which can be obtained by setting

$$\frac{Dv_x}{Dt} = 0, \quad \frac{Dv_y}{Dt} = 0, \quad \frac{Dv_z}{Dt} = 0, \quad \frac{DT}{Dt} = 0 \tag{169}$$

in the Navier–Stokes–Fourier equations. Here $D/Dt = \partial/\partial t + v_x\partial/\partial x + v_y\partial/\partial y + v_z\partial/\partial z$ is the substantive derivative. Therefore, instead of the steady state, our extremum principle in terms of entransy leads to the state that the substantive derivatives of the velocity and temperature fields vanish for incompressible fluid flow. In the sense of Eq. (169), one may say that the extremum principle developed in the present work is compatible with the Navier–Stokes–Fourier equations.

Based on the thermodynamics in terms of entransy, we establish a variational principle for heat conductions which gives rise to the Fourier heat conduction equation at steady state. This principle is further extended and a general variational principle for heat transfer and fluid flow under the incompressible assumption is formulated. We find that this principle leads to the Stokes equations which are the special form of the Navier–Stokes–Fourier equations where the substantive derivatives of the temperature and velocity field vanish. In this sense we conclude that the variational principle in terms of entransy is consistent with the Navier–Stokes–Fourier equations.

7 Application of Entransy Dissipation Theory in Heat Exchanger Design

In this section the expressions of the entransy dissipation related to the irreversibility in heat exchanger are derived. The entransy dissipation number which can be used for evaluating the performance of heat exchangers is defined by a non-dimensionalisation method for the entransy dissipation. The entransy dissipation minimization approach for heat exchanger optimization design is developed.

7.1 Introduction

The heat transfer in heat exchangers usually involves the heat conduction under finite temperature difference, the flow friction under finite pressure drop and fluid mixing. These processes are characterized as irreversible thermodynamic processes. Hence, in recent decades the study of heat exchanger optimization design based on the second law of thermodynamics has attracted a lot of attention [105], culminating at the EGM approach developed by Bejan [12, 13]. However, the entropy generation minimization approach, widely applied to modeling and optimization of thermal systems that owe their thermodynamic imperfection to heat transfer, mass transfer, and fluid flow irreversibilities, demonstrates some inconsistencies and paradoxes [34].

Recently based on the concept of entransy, the heat transfer efficiency and equivalent thermal resistance in a heat exchanger were defined [87]. Wang et al. derived an entransy transfer equation describing the entransy transfer processes of a multi-component viscous fluid subjected to heat transfer by conduction and convection, mass diffusion and chemical reactions [106]. Chen and Ren defined a ratio of temperature difference to heat flux as the generalized thermal resistance of convective heat transfer processes, developed the minimum thermal resistance approach for convective heat transfer optimization, and found that the minimum thermal resistance principle is equivalent to the entransy dissipation extremum principle [107]. Xia et al. studied the optimum parameter distributions in two-fluid flow heat exchanger by using optimal control theory under the fixed heat load condition and taking the entransy dissipation minimization as the optimization objective [108]. Liu et al. investigated the applicability of the extremum principles of entropy generation and entransy dissipation for heat exchanger optimization, and found that the former is better for the heat exchanger optimization when it works in the Brayton cycle, while the latter gives better results when heat exchanger is only for the purpose of heating and cooling [109]. Recently the entransy dissipation extremum principle was extended to the radiative heat transfer by Wu and Liang [110]. Guo et al. introduced a non-dimensionalisation method for the entransy dissipation in heat exchanger and an entransy dissipation number which can be used to evaluate the heat exchanger performance was defined [8].

However, all the work mentioned above is based on the entransy defined by Guo et al. which is only related to the heat conduction in heat exchanger [86]. In Sect. 4 we have re-defined the entransy and entransy dissipation in the framework of thermodynamics which not only describes the ability to transfer heat, but also the ability to perform work. Under this definition the entransy dissipation can quantify not only the irreversibility associated with the heat conduction, but also the irreversibility induced by flow friction.

In this section, our work [7–9, 11] is reviewed. In this work, a dimensionless method for the entransy dissipation of heat exchanger was proposed. Subsequently a quantity called the entransy dissipation number was defined. Guo et al. showed that the entransy dissipation number can be employed to evaluate the heat exchanger performance and avoids 'the entropy generation paradox' [8]. For the water–water balanced counter-flow heat exchanger Li et al. showed that under certain assumptions the duct aspect ratio or mass velocity has opposing effects on the two types of entransy dissipations caused by heat conduction under finite temperature difference and the flow friction under finite pressure drop, therefore, minimizing the overall entransy dissipation in heat exchanger can lead to an optimum duct aspect ratio or mass velocity [11]. Furthermore, the analytic expressions of the optimal duct aspect ratio and mass velocity of the heat exchanger were obtained [11]. Based on the entransy dissipation theory, a principle of entransy dissipation equipartition for heat exchanger optimization design was established [7]. Finally, an entransy dissipation minimization approach for heat exchanger optimization design was developed [9].

7.2 Entransy Dissipation in Heat Exchanger

According to the definition of entransy, we have the following relationship between the entransy and enthalpy

$$dg_0 = Tdh \tag{170}$$

where g_0 and h are the specific entransy and enthalpy, respectively. Assume that the heat exchanger under consideration is an adiabatic open system. By making use of Eq. (170), the entransy dissipation rate induced by heat conduction in the heat exchanger is written as

$$\dot{G}_{diss,1} = \int_{o}^{i} (\dot{m}Tdh)_1 = \int_{o}^{i} (\dot{m}c_pTdT)_1 = \frac{1}{2}(\dot{m}c_p)_1({T_{1,i}}^2 - {T_{1,o}}^2) \tag{171}$$

where $\dot{G}_{diss,1}$ is the entransy dissipation rate of the fluid 1, c_p is the specific heat of fluid at constant pressure and assumed to be a constant, the subscript 1 denotes the fluid 1, the subscripts i and o denote the inlet and outlet, respectively. Similarly, for the fluid 2 we have

$$\dot{G}_{diss,2} = \int_{o}^{i} (\dot{m}Tdh)_2 = \int_{o}^{i} (\dot{m}c_pTdT)_2 = \frac{1}{2}(\dot{m}c_p)_2({T_{2,i}}^2 - {T_{2,o}}^2) \tag{172}$$

Therefore, the total entransy dissipation rate induced by heat conduction in the heat exchanger is expressed as

$$\dot{G}_{diss,T} = \dot{G}_{diss,1} + \dot{G}_{diss,2} = \frac{1}{2}(\dot{m}c_p)_1({T_{1,i}}^2 - {T_{1,o}}^2) + \frac{1}{2}(\dot{m}c_p)_2({T_{2,i}}^2 - {T_{2,o}}^2) \tag{173}$$

Now we consider the entransy dissipation rate caused by flow friction in heat exchangers. Assume that the fluid flow in the heat exchanger is driven by the finite pressure drop between the inlet and outlet, and the flow is stationary and adiabatic. Then from the definition of entransy, we have

$$dg_0 = Tdp/\rho \tag{174}$$

where p is the pressure. Then the entransy dissipation rate induced by flow friction is written as

$$\dot{G}_{diss,p} = -\int_{i}^{-o} \dot{m}dg_0 = -\int_{i}^{o} \dot{m}T/\rho dp \tag{175}$$

If the fluid is the ideal gas, applying its state equation in Eq. (175) gives

$$\dot{G}_{diss,p} = -\int_{i}^{o} \dot{m}T^2R/pdp \tag{176}$$

where R is the ideal gas constant. If replacing the temperature in Eq. (176) with the logarithmic mean temperature, we obtain

$$\dot{G}_{diss,p} = -{T_{lm}}^2 \int_{i}^{o} \dot{m}R/pdp \tag{177}$$

where $T_{lm} = (T_o - T_i)/(\ln T_o - \ln T_i)$. Similarly, for the incompressible fluid we have

$$\dot{G}_{diss,p} = \frac{\dot{m}T_{lm}\Delta p}{\rho} \tag{178}$$

where $\Delta p = p_i - p_o$. If both fluids in the heat exchanger are idea gases, the entransy dissipation rate induced by fluid friction is written as

$$\dot{G}_{diss,p} = -\dot{m}_1\left(\frac{T_{1,o} - T_{1,i}}{\ln T_{1,o} - \ln T_{1,i}}\right)^2 R\ln\frac{p_{1,o}}{p_{1,i}} - \dot{m}_2\left(\frac{T_{2,o} - T_{2,i}}{\ln T_{2,o} - \ln T_{2,i}}\right)^2 R\ln\frac{p_{2,o}}{p_{2,i}} \tag{179}$$

When both fluids in the heat exchanger are incompressible, we obtain

$$\dot{G}_{diss,p} = \frac{\dot{m}_1\Delta p_1}{\rho_1}\frac{T_{1,o} - T_{1,i}}{\ln T_{1,o} - \ln T_{1,i}} + \frac{\dot{m}_2\Delta p_2}{\rho_2}\frac{T_{2,o} - T_{2,i}}{\ln T_{2,o} - \ln T_{2,i}} \tag{180}$$

When the fluid 1 is idea gas, fluid 2 is impressible, we have

$$\dot{G}_{diss,p} = -\dot{m}_1\left(\frac{T_{1,o} - T_{1,i}}{\ln T_{1,o} - \ln T_{1,i}}\right)^2 R\ln\frac{p_{1,o}}{p_{1,i}} + \frac{\dot{m}_2\Delta p_2}{\rho_2}\frac{T_{2,o} - T_{2,i}}{\ln T_{2,o} - \ln T_{2,i}} \tag{181}$$

Finally, the total entransy dissipation rate caused by heat conduction and flow friction in heat exchanger is expressed as

$$\dot{G}_{diss} = \dot{G}_{diss,T} + \dot{G}_{diss,p}. \tag{182}$$

7.3 Entransy Dissipation Number

Assume that both the hot and cold fluids in heat exchanger are incompressible. The inlet temperatures and pressures of the hot and cold fluids are denoted as $T_{1,i}$, $P_{1,i}$ and $T_{2,i}$, $P_{2,i}$ respectively. Similarly the outlet temperatures and pressures are $T_{1,o}$, $P_{1,o}$ and $T_{2,o}$, $P_{2,o}$. For the one-dimensional heat exchanger considered in this section, the usual assumptions such as steady flow, no heat exchange with environment, and ignoring changes in kinetic and potential energies as well as the longitudinal conduction are made. In the heat exchanger, there mainly exist two kinds of irreversibility: the first is heat conduction under finite temperature differences and the second is flow friction under finite pressure drops.

Note that according to the definition of entransy the possible maximum entransy dissipation in a heat exchanger is $Q(T_{1,i} - T_{2,i})$, where Q is the actual heat transfer rate, $(T_{1,i} - T_{2,i})$ is the maximum temperature difference in the heat exchanger. Therefore the entransy dissipation rate caused by heat conduction under a finite temperature difference can be non-dimensionalised in the following way [8]:

$$\dot{G}^*_{diss,T} = \frac{\dot{G}_{diss,T}}{Q(T_{1,i} - T_{2,i})} = \frac{\dot{G}_{diss,T}}{\varepsilon(\dot{m}c)_1 (T_{1,i} - T_{2,i})^2} \tag{183}$$

where ε is the heat exchanger effectiveness which is defined as the ratio of the actual heat transfer rate to the maximum possible heat transfer rate. From Eq. (183) the entransy dissipation number finally arrives at

$$\dot{G}^*_{diss,T} = \frac{2 - \varepsilon(1 + C^*)}{2} \tag{184}$$

where C^* is the heat capacity rate ratio. From Eq. (184), one can see that the entransy dissipation number is dependent on the heat exchanger effectiveness and heat capacity rate ratio, independent on the inlet temperatures of the heat exchanger.

7.3.1 Counter Flow Heat Exchanger

For the counter flow arrangement, the heat exchanger effectiveness is written as [36]

$$\varepsilon = \frac{1 - \exp[-Ntu(1 - C^*)]}{1 - C^* \exp[-Ntu(1 - C^*)]} \tag{185}$$

where *Ntu* is the number of exchanger heat transfer units. Substituting Eq. (185) into Eq. (184) yields

$$\dot{G}^*_{diss,T} = 1 - \frac{1 + C^*}{2} \frac{1 - \exp[-Ntu(1 - C^*)]}{1 - C^* \exp[-Ntu(1 - C^*)]} \tag{186}$$

From Eq. (186) one can see that the entransy dissipation number doesn't tend to zero even when *Ntu* approaches to the infinity. The variation of the entransy dissipation number with the number of exchanger heat transfer units and exchanger effectiveness for various heat capacity rate ratios are depicted in Fig. 30a and b, respectively. Figure 30a shows that the entransy dissipation number monotonously decreases with the increase of the number of exchanger heat transfer units for the fixed heat capacity rate ratio. With the increase of the heat exchanger effectiveness the entransy dissipation number approximately linearly decays, as shown in Fig. 30b. But when the effectiveness reaches its maximum value (one), the entransy dissipation number doesn't vanish. This is evident from the observation of Eq. (184). When $\varepsilon = 1.0$, Eq. (184) becomes

$$\dot{G}^*_{diss,T} = \frac{1 - C^*}{2}$$

Therefore the entransy dissipation number is not equal to zero unless $C^* = 1$ (corresponding to the balanced flow). For the imbalanced flow although the heat

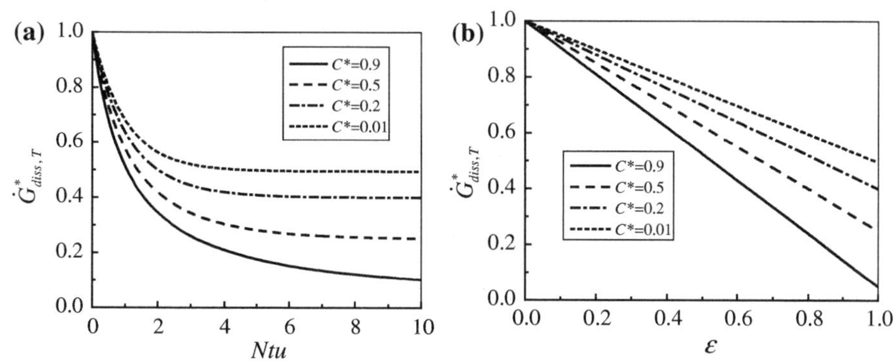

Fig. 30 Variation of the entransy dissipation number for the counter flow heat exchanger with respect to the number of exchanger heat transfer units and exchanger effectiveness. **a** Entransy dissipation number versus *Ntu*. **b** Entransy dissipation number versus ε

transfer area tends to the infinity, the flow imbalance causes irreversibility in the heat exchanger. This kind of irreversibility is called as 'irreversibility due to flow imbalance' discussed by Bejan [13]. Figure 30 shows that the larger the heat capacity rate ratio, the smaller the entransy dissipation number is. Therefore to some extent the entransy dissipation number quantifies the irreversibility caused by the flow imbalance.

Now we turn to the entropy generation number which is defined as the ratio of the entropy generation to the smaller heat capacity rate [12]. When the hot fluid heat capacity rate is smaller, the entropy generation number for the counter flow heat exchanger can be written as follows [34]:

$$N_s = C^* \ln\left[1 - \varepsilon\left(1 - \frac{T_{2,i}}{T_{1,i}}\right)\right] + \ln\left[1 + C^*\varepsilon\left(\frac{T_{1,i}}{T_{2,i}} - 1\right)\right] \tag{187}$$

From Eq. (187), it is evident that the entropy generation number is not only related to the heat exchanger effectiveness and heat capacity rate ratio, but also related to the inlet temperatures of both fluids. When $T_{1,i}/T_{2,i} = 1.25$ the variations of the entropy generation number with the exchanger effectiveness and number of exchanger heat transfer units are displayed in Fig. 31a and b, respectively. From Fig. 31a, one can see that with increasing the exchanger effectiveness, the entropy generation number first increases, then decreases. The dependence of the entropy generation number on the number of heat transfer units demonstrates the similar behavior, as illustrated in Fig. 31b. Therefore there exists a value of the exchanger effectiveness denoted as ε^*, at this point the entropy generation number achieves the maximum. From Eq. (187), we obtain

$$\varepsilon^* = \frac{1}{1 + C^*} \tag{188}$$

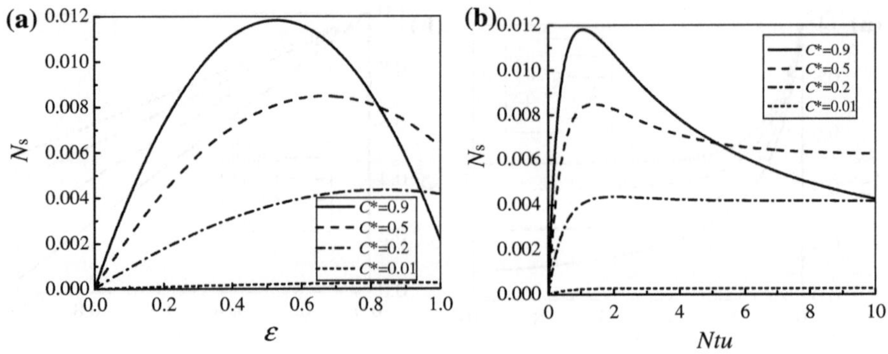

Fig. 31 Variations of the entropy generation number with the heat exchanger effectiveness and number of exchanger heat transfer units. **a** The entropy generation number versus ε ($T_{1,i}/T_{2,i} = 1.25$). **b** The entropy generation number versus *Ntu* ($T_{1,i}/T_{2,i} = 1.25$)

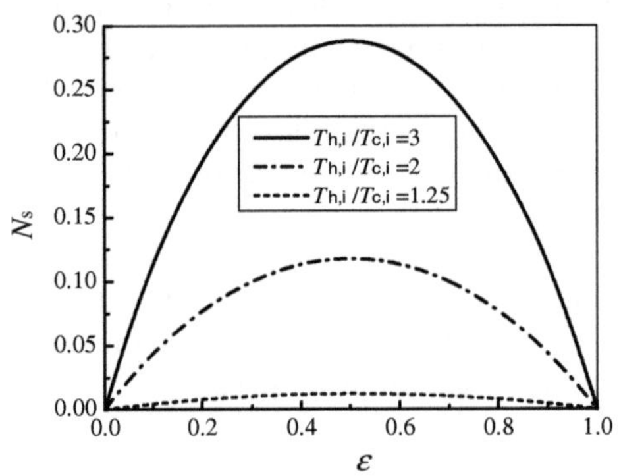

Fig. 32 The dependence of the entropy generation number on the exchanger effectiveness for the balanced counter flow

From Eq. (188), one can see that ε^* is only dependent on the heat capacity rate ratio. When $C^* = 1$, namely the heat capacity rates of the hot and cold flow are same, then according to Eq. (188), $\varepsilon^* = 0.5$. In this case the dependence of the entropy generation number on the exchanger effectiveness is illustrated in Fig. 32. The thermodynamic analysis indicates that the less entropy generation means less thermodynamic irreversibility and better performance of the heat exchanger.

However, from Fig. 32 one can see that the entropy generation number increases with the increase of the heat exchanger effectiveness when the effectiveness is less than ε^*. This phenomenon is called 'entropy generation paradox' [34]. As opposed to the entropy generation number the entransy dissipation number demonstrates the reasonable relationship with the exchanger effectiveness as shown in Fig. 30b.

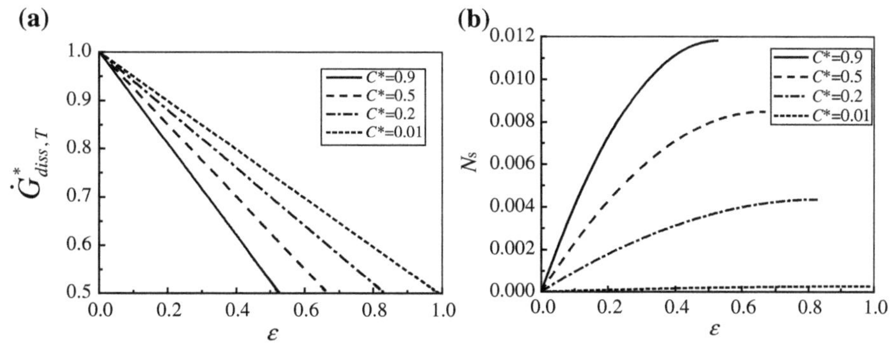

Fig. 33 Variations of the entranspy dissipation number and entropy generation number with the heat exchanger effectiveness for parallel flow heat exchanger ($T_{1,i}/T_{2,i} = 1.25$). **a** The entranspy dissipation number versus the heat exchanger effectiveness. **b** The entropy generation number versus the heat exchanger effectiveness

7.3.2 Parallel Flow Heat Exchanger

For the parallel flow heat exchanger, the heat exchanger effectiveness is written as [36]

$$\varepsilon = \frac{1 - \exp[-Ntu(1 + C^*)]}{1 + C^*} \tag{189}$$

Substituting Eq. (189) into Eq. (184) yields

$$\dot{G}^*_{diss,T} = \frac{1 + \exp[-Ntu(1 + C^*)]}{2} \tag{190}$$

For this case the variation of the entransy dissipation number with the exchanger effectiveness is depicted in Fig. 33a for various heat capacity rate ratios. Note that although the parallel and counter flow heat exchangers share the same formula for the entransy dissipation number, their achievable ranges of the heat exchanger effectiveness are different for the fixed heat capacity rate, as shown in Figs. 30b and 33a. Figure 33b illustrates the variation of the entropy generation number with the exchanger effectiveness for various heat capacity rate ratios. From this figure one can see that the entropy generation number always increases with the increase of the heat exchanger effectiveness. Thus the 'entropy generation paradox' is more prominent than the counter flow heat exchanger. Why does the variation of the entropy generation number with respect to the heat exchanger effectiveness behave so different for these two types of heat exchangers? Actually in the parallel flow heat exchanger the highest possible outlet temperature of the cold fluid is the same as the outlet temperature of the hot fluid in the ideal case. It is in this ideal case that the entropy generation number just reaches its maximum value. For the counter flow heat exchanger when the heat exchanger effectiveness increases and reaches certain value, the outlet temperature of the cold flow equals to the outlet temperature of the hot flow, and the entropy generation number achieves its maximum value. After that if the heat exchanger effectiveness increases further, the outlet temperature of the cold flow will exceeds the outlet temperature of the hot flow, and the entropy generation number begins decay.

7.3.3 Comparison of the Revised Entropy Generation Number with Entransy Dissipation Number

As discussed above, the entropy generation number defined by Eq. (187) suffers from the 'entropy generation paradox'. In an effort to resolve this paradox, a number of ways for non-dimensionalising the entropy generation are proposed, such as by Q/T_{a} (T_{a} is environmental temperature) and $Q/T_{\mathrm{c,i}}$ [34]. In the present work the latter is employed, namely, the entropy generation is non-dimensionalised as follows

$$N_{\mathrm{s1}} = \frac{\dot{S}_{\mathrm{gen}} T_{\mathrm{c,i}}}{Q} \tag{191}$$

N_{s1} is called the revised entropy generation number in the following discussion. The revised entropy generation number avoids the 'entropy generation paradox' and behaves in a more intuitively reasonable way [34]. However, when it is employed to assess the performance of different flow arrangements, some problems occur, as illustrated in Fig. 34. Three curves in Fig. 34 represent the variations of the revised entropy generation numbers of the counter flow arrangement with $T_{1,i}/T_{2,i} = 1.6$, cross flow arrangement with $T_{1,i}/T_{2,i} = 1.5$ and parallel flow arrangement with $T_{1,i}/T_{2,i} = 1.4$ with respect to the number of exchanger heat transfer units. Among these three flow arrangements it is well known that the counter flow arrangement exhibits the best performance [111]. However, this is only consistent with the last parts of three curves where the number of exchanger heat transfer units is large, otherwise the counter flow arrangement is the worst one

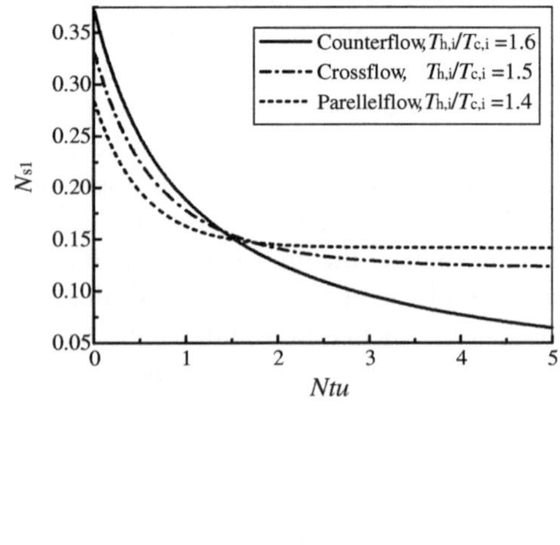

Fig. 34 The revised entropy generation number versus the number of exchanger heat transfer units for different flow arrangements ($C^{*} = 0.9$)

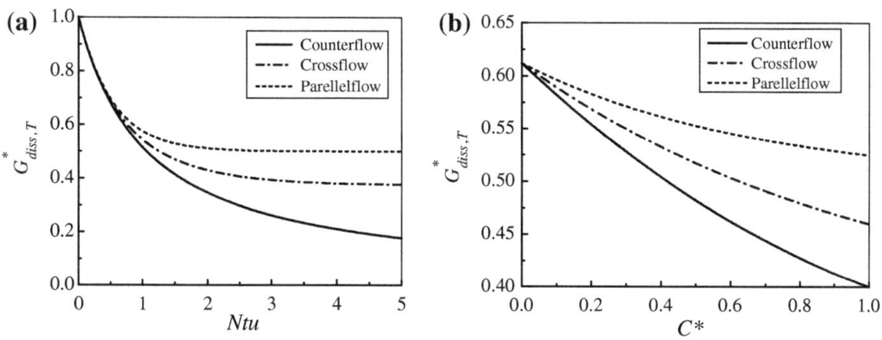

Fig. 35 **a** The entransy dissipation number versus the number of exchanger transfer units ($C^* = 0.9$). **b** The entransy dissipation number versus the heat capacity rate ratio ($Ntu = 1.5$)

in terms of the revised entropy generation number. This is due to the sensitive dependence of the revised entropy generation number on the ratio of the inlet temperature of the hot fluid to the counterpart of the cold fluid. Therefore even the revised entropy generation number is incapable to correctly evaluate the global performance of heat exchangers.

Now let us turn to the entransy dissipation number. The variations of the entransy dissipation number with respect to the number of exchanger heat transfer units and the heat capacity rate ratio are illustrated in Fig. 35a and b, respectively. For both cases, one can clearly see that the counter flow arrangement is the best, the parallel flow arrangement is the worst and the cross flow arrangement is in between. Therefore the entransy dissipation number correctly reflects the global performance of the heat exchangers and can serve as the performance evaluation criterion.

7.3.4 The Overall Entransy Dissipation Number

From Eq. (180), the entransy dissipation rate due to flow friction under a finite pressure drop is expressed as

$$\dot{G}_{diss,P} = \frac{\dot{m}_1 \Delta P_1}{\rho_1} \frac{T_{1,o} - T_{1,i}}{\ln T_{1,o} - \ln T_{1,i}} + \frac{\dot{m}_2 \Delta P_2}{\rho_2} \frac{T_{2,o} - T_{2,i}}{\ln T_{2,o} - \ln T_{2,i}} \tag{192}$$

where ΔP_1 and ΔP_2 refer to the pressure drops of the hot and cold water, respectively; ρ_1 and ρ_2 are their corresponding densities. Putting in dimensionless form leads to

$$\dot{G}^{*}_{diss,P} = \frac{-\Delta P_1}{(\rho c)_1 (T_{1,i} - T_{2,i})} \frac{1}{\ln\frac{T_{2,i}}{T_{1,i}}\left[1 - (1-\varepsilon)\frac{T_{2,i}-T_{1,i}}{T_2}\right]} + \frac{\Delta P_2}{(\rho c)_2 (T_{1,i} - T_{2,i})} \frac{1}{\ln\frac{T_{1,i}}{T_{2,i}}\left[1 + (1-\varepsilon)\frac{T_{2,i}-T_{1,i}}{T_{1,i}}\right]} \tag{193}$$

which is called the entransy dissipation number due to flow friction.

Assuming that the heat exchanger behaves as a nearly ideal heat exchanger, then $(1-\varepsilon)$ is considerably smaller than unity [12]. For a water–water heat exchanger under usual operating conditions, the inlet temperature difference between hot and cold water, $\Delta T = T_{1,i} - T_{2,i}$, is less than 100 K, hence $\Delta T / T_i < 100/273 \approx 0.366$ $(i = 1, 2)$. Therefore, Eq. (193) can be simplified to

$$\dot{G}^{*}_{diss,P} = \frac{\Delta P_1}{(\rho c)_1 (T_{1,i} - T_{2,i})} \frac{1}{\ln\frac{T_{1,i}}{T_{2,i}}} + \frac{\Delta P_2}{(\rho c)_2 (T_{1,i} - T_{2,i})} \frac{1}{\ln\frac{T_{1,i}}{T_{2,i}}} \tag{194}$$

Accordingly, the overall entransy dissipation number becomes

$$\begin{aligned} \dot{G}^{*}_{diss} &= \dot{G}^{*}_{diss,T} + \dot{G}^{*}_{diss,P} \\ &= (1-\varepsilon) + \frac{\Delta P_1}{(\rho c)_1 (T_{1,i} - T_{2,i})} \frac{1}{\ln\frac{T_{1,i}}{T_{2,i}}} + \frac{\Delta P_2}{(\rho c)_2 (T_{1,i} - T_{2,i})} \frac{1}{\ln\frac{T_{1,i}}{T_{2,i}}} \end{aligned} \tag{195}$$

For a typical water–water balanced heat exchanger, the number of heat transfer units *Ntu* can be introduced, which approaches infinity as the effectiveness tends to unity. For the balanced heat exchanger, the heat capacity rate ratio satisfies condition $C^* = (\dot{m}c)_2/(\dot{m}c)_1 = 1$ (where $\dot{m}$ is the mass flow rate), therefore the effectiveness is [12]:

$$\varepsilon = \frac{Ntu}{1 + Ntu} \tag{196}$$

where the number of heat transfer units is defined as

$$Ntu = \frac{UA}{\dot{m}c_p}$$

here U is the overall heat transfer coefficient, and A is the heat transfer area. Assume that compared with the convective heat transfer the heat conduction resistance of the solid wall can be neglected. It is appropriate to replace U with the convective heat transfer coefficient h. Therefore,

$$\frac{1}{UA} = \frac{1}{(hA)_1} + \frac{1}{(hA)_2} \tag{197a}$$

or

$$\frac{1}{Ntu} = \frac{1}{Ntu_1} + \frac{1}{Ntu_2} \tag{197b}$$

where h_1 and h_2 are the convective heat transfer coefficients of the hot and cold fluids, respectively, and $Ntu_i = (hA)_i/(\dot{m}c)_i (i = 1, 2)$. In the nearly ideal heat exchanger limit, $Ntu \gg 1$, that is [12]

$$1 - \varepsilon \approx \frac{1}{Ntu} \tag{198}$$

Therefore, the overall entransy dissipation number becomes

$$\dot{G}^* = \left[\frac{1}{Ntu_1} + \frac{\Delta P_1}{(\rho c)_1 (T_{1,i} - T_{2,i}) \ln\frac{T_{1,i}}{T_{2,i}}} \frac{1}{} \right] + \left[\frac{1}{Ntu_2} + \frac{\Delta P_2}{(\rho c)_2 (T_{1,i} - T_{2,i}) \ln\frac{T_{1,i}}{T_{2,i}}} \frac{1}{} \right] \tag{199}$$

The two terms on the right of Eq. (199) correspond to the entransy dissipations of two sides of heat transfer surfaces. For each side, the entransy dissipation number can be written as follows:

$$\dot{G}^*_{diss,i} = \frac{1}{Ntu_i} + \frac{P_i}{(\rho c)_i (T_{1,i} - T_{2,i}) \ln\frac{T_{1,i}}{T_{2,i}}} \frac{\Delta P_i}{P_i} \quad (i = 1, 2) \tag{200}$$

It is evident that the first term accounts for the entransy dissipation from the heat conduction under finite temperature difference and the second for the entransy dissipation from flow friction under finite pressure drop. For simplicity, we now use $\dot{G}^*_{diss}$ instead of $\dot{G}^*_i$ to denote the entransy dissipation number for each side of the heat exchanger surface.

7.4 Parameter Optimization

Theoretically, the exchanger effectiveness increases when the irreversible dissipation in the heat exchanger decreases. From Sect. 6.3 one can see that the entransy dissipation can be used for describing the irreversible dissipations, therefore we seek optimums in duct aspect ratio and mass velocity by minimizing the entransy dissipation number $\dot{G}^*_{diss}$ based on Eq. (200). Although the entransy dissipation number on one side of the heat transfer surface can be expressed as the sum of the contributions of the heat conduction under the finite temperature difference and flow friction under the finite pressure drop, the effects of these two factors on heat exchanger irreversibility are strongly coupled through the geometric parameters of the heat exchanger tube residing on that side. Therefore, based on entransy dissipation minimization, it is possible to obtain optimal geometric parameters of the heat exchanger such as the optimal duct aspect ratio.

Recall the definition of the Stanton number $St((\mathrm{Re})_D, \mathrm{Pr})$ and friction factor $f((\mathrm{Re})_D)$:

$$Ntu = \frac{4L}{D} St \tag{201}$$

$$\frac{\Delta P}{P} = f \frac{4L}{D} \frac{G^2}{2\rho P} \tag{202}$$

where $G = \dot{m}/A$ is the mass velocity, L is the flow path length and D is the duct hydraulic diameter. Introducing the dimensionless mass velocity, $G_* = G/\sqrt{2\rho P}$

$$\tau^2 = \frac{P}{(\rho c)(T_{1,i} - T_{2,i}) \ln \frac{T_{1,i}}{T_{2,i}}}$$

and substituting Eqs. (201) and (202) into Eq. (200), we obtain

$$\dot{G}^*_{diss} = \frac{1}{\frac{4L}{D} St} + \tau^2 f \frac{4L}{D} G_*{}^2 \tag{203}$$

Clearly, the duct aspect ratio $4L/D$ has opposing effects on the two terms of the right side of Eq. (203). Therefore, there exists an optimal duct aspect ratio to minimize the entransy number. When the Reynolds number and mass velocity are fixed, minimizing the entransy dissipation number leads to the following expression:

$$\left(\frac{4L}{D}\right)_{opt} = \frac{1}{\tau G_* (fSt)^{1/2}} \tag{204}$$

The corresponding minimum entransy dissipation number is

$$\dot{G}^*_{diss,\min} = 2\tau G_* \left(\frac{f}{St}\right)^{1/2} \tag{205}$$

From Eqs. (204) and (205), one can see that the optimal duct aspect ratio decreases as the mass velocity G_* increases, and the minimum entransy dissipation number is directly proportional to the dimensionless mass velocity. Note that the minimum entransy dissipation number is also dependent on the Reynolds number via f and St. However, the impact of the Reynolds number on the minimum entransy dissipation number is very weak since for many heat transfer surfaces the ratio of the friction factor to the Stanton number does not have a significant change as the Reynolds number varies [12]. Therefore, the minimum entransy dissipation number is mainly determined by the selected dimensionless mass velocity. Obviously, the smaller the mass velocity, the longer the working fluid remains on the heat transfer surface and the lower the irreversible dissipations in the heat exchanger. In designing a heat exchanger, the heat transfer area is an important consideration when it accounts for most of the total cost of a heat exchanger. Thus in the following, we discuss design optimization of the heat exchanger with the fixed heat transfer area.

From the definition of the hydraulic diameter, the heat transfer area for one side is

$$A = \frac{4L}{D} A_c$$

where A_c is the duct cross-section area. This expression can be put in dimensionless form as

$$A_* = \frac{4L}{D} G_*^{-1} \tag{206}$$

where A_* is the dimensionless heat transfer area and $A_* = (2\rho P)^{1/2} A/\dot{m}$. Substituting Eq. (206) into Eq. (203) yields

$$\dot{G}^*_{diss} = \frac{1}{A_* St} G_*^{-1} + \tau^2 f A_* G_*^3 \tag{207}$$

Obviously, the dimensionless mass velocity has an opposing effect on the two terms of the right side of Eq. (207). Thus there exists an optimal dimensionless mass velocity which allows the entransy dissipation number to reach a minimum value when A_* and Reynolds number $(\mathrm{Re})_D$ are given. Solving this optimization problem yields:

$$G_{*,opt} = \left(\frac{1}{3 A_*^2 \tau^2 f St} \right)^{1/4} \tag{208}$$

$$\dot{G}^*_{diss,\min} = 4 \left(\frac{\tau^2 f}{27 A_*^2 St^3} \right)^{1/4} \tag{209}$$

Equations (208) and (209) respectively give the optimal dimensionless mass velocity and the minimum entransy dissipation number under the fixed A_* and Reynolds number $(\mathrm{Re})_D$. From these two equations, one can see that the larger heat transfer area clearly corresponds to the smaller mass velocity and lower entransy dissipation rates. Hence, to reduce the irreversible dissipation occurring in a heat exchanger, the largest-possible heat transfer area should be adopted under the allowable conditions.

If $\dot{G}^*_{diss}$ and $(\mathrm{Re})_D$ are given, the minimum heat transfer area is

$$A_{*,\min} = \frac{16}{3^{3/2}} \frac{\tau f^{1/2}}{\dot{G}^{*2}_{diss} St^{3/2}} \tag{210}$$

with

$$\left(\frac{4L}{D} \right)_{opt} = \frac{4}{3} \frac{1}{\dot{G}^*_{diss} St} \tag{211}$$

From Eqs. (210) and (211), one can see that a lower entransy dissipation rate corresponds to a larger heat transfer area or duct aspect ratio. Equations (210) and (211) are identical, providing an expression for the minimum attainable value for the product $A_*^{1/2}\dot{E}^*$ under the given Reynolds number.

As exemplified by the water–water counter-flow heat exchanger, we show that there exists an optimal duct aspect ratio for heat exchangers under the fixed Reynolds number and mass velocity when the entransy dissipation number is taken as the performance evaluation criterion. Furthermore, the formula for the optimal duct aspect ratio was obtained analytically. Under constraints of the fixed heat transfer area (or duct volume) and Reynolds number, it was shown that there is an optimal dimensionless mass velocity; for which an analytical expression was also given. The results indicated that to reduce irreversible dissipations in heat exchangers, largest-possible heat transfer areas and lowest-possible mass velocities should be adopted.

7.5 Entransy Dissipation Minimization Principle

7.5.1 Introduction

In recent decades, the application of the second law of thermodynamics in heat exchangers has aroused a lot of attention [105]. The heat transfer in heat exchange process usually involves the heat conduction under finite temperature difference and the flow friction. They are the typical irreversible non-equilibrium thermodynamic processes. Based on the principle of entropy production minimization advanced by Wang et al. [106], Bejan [12, 13] developed the entropy generation minimization approach to the heat exchanger optimization design. Tondeur and Kvaalen found that the total entropy production reaches the minimum when the local rate of entropy production is uniformly distributed along the space and/or time variables in a contacting or separation device involving a given transfer area and achieving a specified transfer duty [112], this principle is called equipartition of entropy production (EoEP). Sauar et al. showed that the best trade-off between energy dissipation and transfer area is achieved when the thermodynamic driving forces are uniformly distributed over the heat transfer area, which is called equipartition of forces (EoF) [113]. Balkan [114] revealed that entropy production calculated with the EoEP is always smaller than those calculated with the EoF although the difference is considerably small. Guo et al. showed that the better uniform distribution of temperature difference along the counter flow heat exchanger contributes the better performance of counter flow exchanger in comparison with the parallel flow type [115]. A principle of the equipartition of temperature difference, EoTD, was proposed as a short-cut way for minimizing entropy production by Balkan [114].

Although the entropy generation minimization is widely applied in heat exchanger design, it also causes some inconsistencies and paradoxes [34, 61, 84]. Guo et al. defined a new physical quantity—entransy, which describes the heat transfer ability. It was found that the entransy is dissipated in irreversible heat conduction processes [86, 116]; the more dissipation of entransy indicates the higher degree of irreversibility [117]. Thus, the entransy dissipation may serve as a figure of merit for assessing the irreversibilities of heat transfer processes. Xu found that the entransy is a state variable and can be used to describe the second law of thermodynamics [4]. Liu et al. showed that the extremum principle of entransy dissipation is advantageous over the extremum principle of entropy generation when heat exchanger is only for heating and cooling, while the latter is better than the former when the heat exchanger works in the Brayton cycle [109]. Wu and Liang extended the entransy dissipation extremum principle to the radiative heat transfer optimization [110]. The principle of entransy dissipation minimization was further applied to constructal optimization for geometry of cavity [108], and volume-point conduction optimization [118]. Recently, based on the entransy dissipation minimization, Wei et al. carried out the constructal multidisciplinary optimization of electromagnet [119].

In this subsection, the principles for minimizing the entransy dissipation due to heat conduction are discussed for heat exchanger design with the given heat duty and heat transfer area, and the differences between entropy generation minimization principle and entransy dissipation minimization principle are investigated.

7.5.2 Principle of Entransy Dissipation Equipartition

Consider a simple heat exchange process as shown in Fig. 36, a thin metal plate separates the hot and cold fluids. Assume both fluids are perfectly mixed in y-direction. The length of the heat exchanger is l, heat is only conducted in the x-direction, and the bulk temperatures of both fluids only vary in z-direction. In this simple heat exchanger, the condition for the minimum entropy generation has been given in [120], herein we try to find the condition of minimum entransy dissipation rate for a given heat duty $\dot{Q}$ (the parameters of hot fluid are fixed) through the similar analysis. According to the expression of entransy dissipation given in [13], the local entransy dissipation rate in the heat exchanger can be expressed as

Fig. 36 A simple heat exchange system

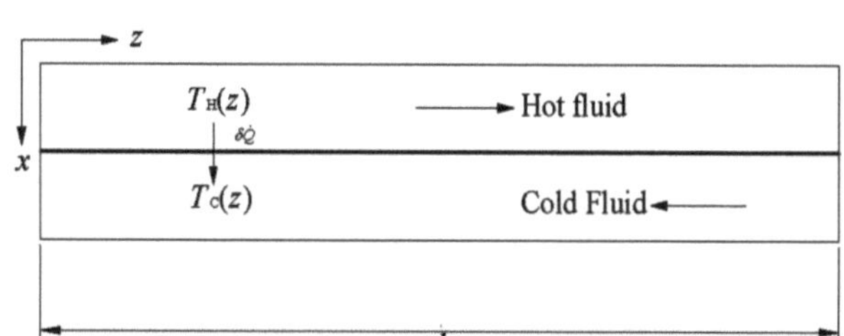

$$\phi = -J_q \frac{\mathrm{d}T}{\mathrm{d}x} \tag{212}$$

where T is the temperature, dT/dx is the driving force, and J_q is the heat flux in x-direction, and is kept constant. When the temperature difference ΔT is small, Eq. (212) can also be written as

$$\phi = -\frac{1}{\Delta x} J_q \Delta T \tag{213}$$

The driving force is [116]

$$X = -\Delta T = RJ_q \tag{214}$$

where R is the thermal resistance. Therefore, Eq. (213) becomes

$$\phi = \frac{1}{\Delta x} R\left(J_q\right)^2 \tag{215}$$

The total entransy dissipation rate can be obtained by integrating ϕ over the whole volume of the exchanger

$$\dot{G}_{\mathrm{diss}} = \Delta y \int_0^l R\left(J_q\right)^2 \mathrm{d}z \tag{216}$$

The local energy balance in the heat exchanger is written as follows

$$\dot{m}_{\mathrm{h}} c_{p\mathrm{h}} \mathrm{d}T_{\mathrm{h}} = -J_q \Delta y \mathrm{d}z \tag{217}$$

where $\dot{m}_{\mathrm{h}}$ is the mass flow rate of the hot fluid, c_{ph} is the specific heat of the hot fluid. Thus, the heat flux can be expressed as

$$J_q = -\frac{\dot{m}_{\mathrm{h}} c_{p\mathrm{h}}}{\Delta y} \frac{\mathrm{d}T_{\mathrm{h}}}{\mathrm{d}z} \tag{218}$$

Substituting Eq. (218) into Eq. (216) yields

$$\dot{G}_{\mathrm{diss}} = \int_0^l \frac{R}{\Delta y} \left(\dot{m}_{\mathrm{h}} c_{p\mathrm{h}} \frac{\mathrm{d}T_{\mathrm{h}}}{\mathrm{d}z}\right)^2 \mathrm{d}z \tag{219}$$

According to variational method, when $\dot{G}_{\mathrm{diss}}$ reaches the minimum under the condition that the total heat transfer rate is constant, the integrand in Eq. (219) must satisfy

$$L - \frac{\mathrm{d}T_{\mathrm{h}}}{\mathrm{d}z} \frac{\partial L}{\partial(\mathrm{d}T_{\mathrm{h}}/\mathrm{d}z)} = C_1 \tag{220}$$

where L stands for the integrand in Eq. (219), C_1 is constant. Eq. (220) can be written as follows

$$\frac{R}{\Delta y}\left(\dot{m}_{\mathrm{h}}c_{p\mathrm{h}}\frac{\mathrm{d}T_{\mathrm{h}}}{\mathrm{d}z}\right)^2-\frac{\mathrm{d}T_{\mathrm{h}}}{\mathrm{d}z}\left[\frac{2R}{\Delta y}\left(\dot{m}_{\mathrm{h}}c_{p\mathrm{h}}\right)^2\frac{\mathrm{d}T_{\mathrm{h}}}{\mathrm{d}z}\right]=-\frac{R}{\Delta y}\left(\dot{m}_{\mathrm{h}}c_{p\mathrm{h}}\frac{\mathrm{d}T_{\mathrm{h}}}{\mathrm{d}z}\right)^2 \\ =-\Delta x\Delta y\phi=C_1 \tag{221}$$

Equation (221) implies that the local entransy dissipation rate is uniformly distributed along the heat exchanger when the total entransy dissipation rate achieves the minimum.

When minimizing $\dot{G}_{\mathrm{diss}}$ we should apply the condition that the overall heat transfer rate is constant

$$\dot{Q}=\Delta y\int_0^l J_q\mathrm{d}z=\text{constant} \tag{222}$$

In order to incorporate this constraint condition, we introduce the Lagrange multiplier λ. Then the application of the Euler–Lagrange method yields

$$\frac{\delta}{\delta J_q}\int\left(R\left(J_q\right)^2+\lambda J_q\right)\mathrm{d}z=0 \tag{223}$$

Therefore,

$$RJ_q=X=-\frac{\lambda}{2}. \tag{224}$$

This equation indicates that the temperature difference is uniformly distributed over the heat transfer area when the total entransy dissipation reaches the minimum. This is called the principle of equipartition of temperature difference (EoTD). In deriving Eq. (224), the assumption that the resistance R is independent of J_q has been applied. However, the reality is that R depends on J_q since the resistance R is related to the amount of heat that has been transferred up to the point z [120], which is expressed as

$$q=\Delta y\int_0^z J_q dz' \tag{225}$$

where $z\leq l$. The solution of the variational problem described by Eqs. (223) and (225) is written as [121]

$$X=-\frac{\lambda}{2}-\frac{1}{2}\int_z^l\frac{\mathrm{d}R}{\mathrm{d}q}\Delta y\left(J_q\right)^2\mathrm{d}z' \tag{226}$$

The derivative of X with respect to z is [120]

$$\frac{dX}{dz} = \frac{1}{2}\frac{dR}{dz}J_q \tag{227}$$

According to Eq. (214), dX/dz is expressed as follows

$$\frac{dX}{dz} = \frac{dR}{dz}J_q + R\frac{dJ_q}{dz} \tag{228}$$

From Eqs. (227) and (228), we obtain

$$\frac{d}{dz}\left(R(J_q)^2\right) = 0 \tag{229}$$

Therefore

$$\frac{d}{dz}\phi = 0. \tag{230}$$

Equation (230) implies that the principle of equipartition of entransy dissipation is still valid when $dR/dR \neq 0$. Under the condition that the heat duty and heat transfer area are constants, the EoED principle is equivalent to the EoTD principle when the resistance R is fixed (or heat transfer coefficient is fixed).

7.5.3 Comparison of EoED with EoTD

In this section, the heat exchanger as shown in Fig. 36 will be optimized by the EoED and EoTD principles, respectively. The results obtained by these two methods will be compared with each other. The known data for the heat exchanger under consideration are documented in Table 18. The heat duty is fixed since the mass flow rate, the inlet and outlet temperatures of hot fluid are given. The minimum total entransy dissipation is achieved only by changing the parameters of cold fluid. Here, water is selected as the working fluid, and its specific heat is expressed as the cubic function of temperature. The heat transfer coefficient is written as follows

$$U = T_{\mathrm{h}}^2/200 \tag{231}$$

Table 18 The known data for heat exchanger with the unfixed heat transfer coefficient

Item	Value
Mass flow rate of hot fluid (kg/s)	1
Inlet temperature of hot fluid (K)	370
Outlet temperature of hot fluid (K)	320
The length of heat exchanger in x-direction (m)	0.1
The length of heat exchanger in y-direction (m)	0.1
The length of heat exchanger (m)	100

The resistance R is the reciprocal of the heat transfer coefficient. Based on the EoED principle, the following expression can be obtained from Eq. (221)

$$\frac{dT_h}{dz} = -\frac{\sqrt{-\Delta y C_1}}{\dot{m}_h c_{ph} \sqrt{R}} \tag{232}$$

When the inlet and outlet temperatures of hot fluid are fixed, C_1 can be solved from Eq. (232) numerically, which gives rise to the temperature profile of the hot fluid in the heat exchanger. The local entransy dissipation rate can be obtained from Eq. (221)

$$\phi = -\frac{C_1}{\Delta x \Delta y} \tag{233}$$

By the EoTD principle and Eqs. (214) and (217), we have

$$\frac{dT_h}{dz} = \frac{\Delta y \Delta T}{\dot{m}_h c_{ph} R} \tag{234}$$

From Eqs. (214) and (224) we have

$$\frac{dT_h}{dz} = -\frac{\Delta y \lambda}{2 \dot{m}_h c_{ph} R} \tag{235}$$

From Eq. (235) λ can be determined numerically when the inlet and outlet temperatures of the hot fluid are given. Accordingly the temperature profile of the hot fluid in the heat exchanger is determined. From Eqs. (214) and (215), the local entransy dissipation rate can be written as

$$\phi = \frac{\lambda^2}{4 \Delta x R} \tag{236}$$

The total entransy dissipation rate is obtained by integrating Eq. (236) over the whole domain. The results obtained by the EoED and EoTD principles are documented in Table 19. In this table $\dot{m}c$ is the mass flow rate of the cold fluid, and ε represents the exchanger effectiveness. From Table 19, one can see that the effectiveness obtained by the EoED principle is larger than that obtained by the EoTD principle. While the total entransy dissipation rate and the mass flow rate of the cold fluid obtained by the EoED principle are less than that obtained by the EoTD principle. This indicates that the optimal result obtained by the EoED principle is better than that obtained by the EoTD principle.

Table 19 The optimization results obtained by the EoED and EoTD principles

Item	EoED	EoTD
$\dot{G}$diss (W K)	7,414,119	7,425,069
$\dot{m}$c (kg/s)	0.9085	1.0023
ε	0.6254	0.5858

7.5.4 Comparison of EoED with EoTD

For the heat exchange process as shown in Fig. 37, if P is the perimeter for heat exchange surface, U is the heat transfer coefficient and remain unchanged, the differential heat at a point z can be written as

$$\delta \dot{Q} = U(P\mathrm{d}z)(T_{\mathrm{H}}(z) - T_{C}(z)) \tag{237}$$

where the subscripts H and C represent the hot and cold fluids, respectively. The total entransy dissipation rate is expressed as

$$\begin{aligned} \dot{G}_{\mathrm{diss}} &= \int_0^l UP(T_{\mathrm{H}}(z) - T_{C}(z))^2 \mathrm{d}z \\ &= UP \int_0^l (T_{\mathrm{H}}(z) - T_{C}(z))^2 \mathrm{d}z \end{aligned} \tag{238}$$

By denoting $(T_{\mathrm{H}}(z) - T_{C}(z))$ as Z, the optimization problem is formulated as follows

$$\begin{aligned} \text{minimize} \quad & \dot{G}_{\mathrm{diss}} = UP \textstyle\int_0^l Z^2 \mathrm{d}z \\ \text{with} \quad & \dot{Q} = UP \textstyle\int_0^l Z\mathrm{d}z = \text{constant} \end{aligned} \tag{239}$$

By applying the Euler–Lagrange method, we obtain

$$2Z + \lambda = 0 \tag{240}$$

That is

$$T_{\mathrm{H}}(z) - T_{C}(z) = \text{constant} \tag{241}$$

Thus the same result as Eq. (224) is obtained, i.e., the EoED principle is equivalent to the EoTD principle when the heat transfer coefficient is fixed.

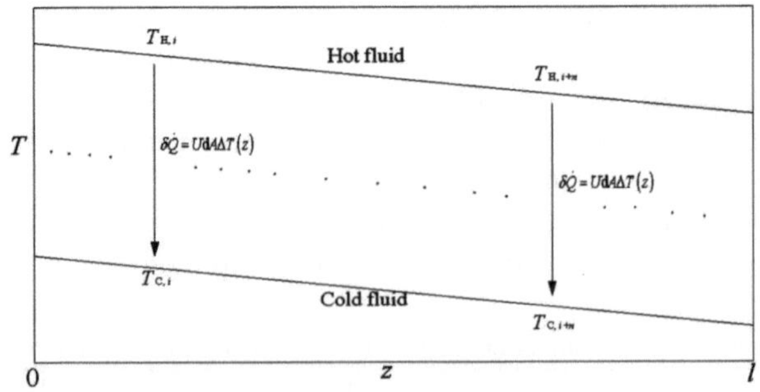

Fig. 37 Heat exchange process without heat-work conversion

7.5.5 Examples for Entropy Generation Minimization and Entransy Dissipation Minimization

Dimensionless entropy generation was first defined as $N_s = \dot{S}_{gen}/(\dot{m}c_p)$ by Bejan [16], N_s is usually called entropy generation number. Unfortunately, it is found that the definition may lead to some 'entropy generation paradoxes' [61]. In an attempt to eliminate this kind of paradoxes, Hesselgreaves proposed the following way to non-dimensionalise the entropy generation [34]

$$N_{s1} = \frac{\dot{S}_{gen} T_{c,i}}{\dot{Q}} \quad (242)$$

where $\dot{S}_{gen}$ is the total entropy generation rate, $T_{c,i}$ is the inlet temperature of the cold fluid. In the following discussion, N_{s1} is employed to quantify the irreversibility in heat exchange process for the entropy generation minimization approach.

In Ref. [8], we introduce the following way to non-dimensionalise the entransy dissipation rate

$$\dot{G}^*_{diss} = \frac{\dot{G}_{diss}}{\dot{Q}(T_{h,i} - T_{c,i})} \quad (243)$$

where $T_{h,i}$ is the inlet temperature of the hot fluid and $\dot{G}^*_{diss}$ is called the entransy dissipation number. This number represents the ratio of the actual entransy dissipation rate to the possible maximum entransy dissipation rate in heat exchanger. From Eq. (243), we obtain

$$\dot{G}^*_{diss} = \frac{2 - \varepsilon(1 + C^*)}{2} \quad (244)$$

where C^* is the heat capacity flow rate ratio, ε is exchanger effectiveness.

For a heat exchanger as shown in Fig. 36, its known data are documented in Table 20. In the following, the total entropy generation rate, the total entransy dissipation rate and other parameters are calculated by the EoEP principle, the EoF principle and the EoED principle, respectively. In each case, the inlet and outlet temperatures of the hot fluid remain unchanged, i.e., the total heat transfer rate in this system is kept constant. Thus, the minimum entropy generation rate and entransy dissipation rate can only be achieved by changing the parameters of the

Table 20 The known data for heat exchanger with the fixed heat transfer coefficient

Item	Value
Total heat transfer coefficient (W/(m^2 K))	800
Mass flow rate of hot fluid (kg/s)	0.2
Specific heat of hot fluid (J/(kg K))	4,181.7
Heat transfer area (m^2)	0.7
Specific heat of cold fluid (J/(kg K))	4,181.7

Table 21 The results obtained by the EoEP, EoF and EoED principles

	$\dot{m}c$ (kg/s)	$T_{c,i}$ (K)	$T_{c,o}$ (K)	$\dot{S}gen$ (W/K)	N_{s1}	$\dot{E}diss$ (W K)	E^*
Case 1: $T_{h,i} = 343.15$ K, $T_{h,o} = 333.15$ K							
EoEP	0.2092	318.44	327.99	1.1430	0.0436	124,910	0.6043
EoF	0.2189	318.64	327.78	1.1431	0.0436	124,940	0.6096
EoED	0.2000	318.22	328.22	1.1431	0.0435	124,900	**0.5990**
Case 2: $T_{h,i} = 353.15$ K, $T_{h,o} = 333.15$ K							
EoEP	0.2191	304.14	322.40	4.6503	0.0846	499,760	0.6097
EoF	0.2400	304.92	321.58	4.6516	0.0848	500,140	0.6199
EoED	0.2000	303.28	323.28	4.6518	0.0843	499,620	**0.5990**
Case 3: $T_{h,i} = 363.15$ K, $T_{h,o} = 333.15$ K							
EoEP	0.2296	290.25	316.39	10.6584	0.1233	1,124,800	0.6150
EoF	0.2635	291.86	314.63	10.6650	0.1241	1,126,600	0.6299
EoED	0.2000	288.35	318.35	10.6671	0.1226	1,124,100	**0.5990**

cold fluid. The calculation results are listed in Table 21. In this table, $\dot{m}_c$ represents the mass flow rate of the cold fluid, $T_{h,o}$ and $T_{c,o}$ are the outlet temperatures of the hot and cold fluids, respectively. From Table 21, one can see that the result obtained by the EoED is best, which is not only reflected by the entransy dissipation number but also by N_{s1}. The optimization results are also in accordance with the conclusions given in [114]. The total entropy generation rate calculated with the EoED is the maximum one. The total entropy generation rate calculated with the EoF is in between. For the entransy dissipation, the total entransy dissipation rate obtained by the EoED is the minimum one, the total entransy dissipation rate calculated with the EoF is the maximum one, and the total entransy dissipation rate calculated with the EoEP is in between. This is consistent with the deduction that the total entransy dissipation rate reaches the minimum when the driving force—temperature difference, is uniformly distributed along the heat exchanger for the given heat duty and heat transfer area. For three cases, N_{s1} and $\dot{G}^*_{diss}$ demonstrate the same tendency. This tendency is also in accordance with the variation of the mass flow rate of the cold fluid. In Table 21, contrary to what is expected, N_{s1} calculated with the EoEP is not the minimum one. While N_{s1} calculated with the EoED reaches the minimum. Therefore N_{s1} does not faithfully represent the trend of entropy generation rate. Note that the entransy dissipation number has the same trend as the total entransy dissipation rate in every case. More importantly, the entransy dissipation number obtained by the EoED remains unchanged when the inlet and outlet temperatures of the hot fluid experience the changes indicated in Table 21. Thus the entransy dissipation number is not directly dependent on the inlet and outlet conditions of heat exchanger. Therefore, the dissipation number is more suitable for evaluating the performance of heat exchanger.

In summary, when local entransy dissipation rate is uniformly distributed along the heat exchanger, the total entransy dissipation rate reaches the minimum under the condition that the heat duty and heat transfer area are fixed. When the heat transfer coefficient is unfixed, the total entransy dissipation rate and entropy generation number obtained by the EoED principle are less than that obtained by the EoTD principle, while the exchanger effectiveness obtained by former is larger than that obtained by the latter. Therefore, the EoED principle demonstrates some advantages over the EoEP and EoF. When the heat transfer coefficient is a constant, the EoED principle is equivalent to the EoTD principle. We find that the entropy generation number is directly related to the inlet and outlet temperatures of working fluids. Consequently, it is unsuitable to be used as a criterion to evaluate the performances of different heat exchangers. While the entransy dissipation number does not directly rely on the inlet and outlet temperatures of working fluid, so it is more suitable for evaluating the performance of different heat exchangers.

7.6 Application of Entransy Dissipation Minimization Approach in Heat Exchanger Design

The heat conduction and fluid friction are two main detrimental irreversibilities in heat exchanger. According to the entransy dissipation theory, the entransy dissipation can be employed to quantify these two irreversibilities. In this subsection, based on the entransy dissipation theory and genetic algorithm an optimization method of heat exchanger design is proposed. Firstly, by taking the total dimensionless entransy dissipation related to heat conduction and fluid friction in heat exchanger as the objective function, a single-objective optimization approach to heat exchanger design is developed. However, it is found that the role played by the fluid friction is not fully taken into account in this approach when the liquid is used for the working fluid. In order to circumvent this problem, the dimensionless entransy dissipations associated with heat conduction and fluid friction are taken as two separate objective functions and a multi-objective optimization approach to heat exchanger design is established. In comparison with the single-objective optimization approach, the multi-objective optimization approach demonstrates more advantages and flexibilities for heat exchanger design.

7.6.1 Introduction

The heat transfer in heat exchangers usually involves the heat conduction under finite temperature difference and the fluid friction under finite pressure drop which are characterized as irreversible non-equilibrium thermodynamic processes. Therefore reducing the irreversibility in heat exchange process is one of powerful measures for improving the heat exchanger performance. Recent decades witness the wide application of the second law thermodynamics in optimization designs of heat exchangers [105]. Inspired by the minimum entropy production principle advanced by Prigogine [59], Bejan [12, 13] developed the entropy generation minimization approach for heat exchanger optimization design. In this approach,

Bejan [12] took into account two types of the irreversibilities in heat exchanger, namely, the heat conduction under the stream-to-stream temperature difference and the frictional pressure drop that accompanies the circulation of fluid through the apparatus. However, the entropy generation minimization approach, widely applied to modeling and optimization of thermal systems that owe their thermodynamic imperfection to heat transfer, mass transfer, and fluid flow irreversibilities, demonstrates some inconsistencies and paradoxes [34]. In an attempt to improve the entropy generation minimization approach Guo et al. [86, 116] defined a new physical concept, entransy, which describes the heat transfer ability. Based on the entransy, the heat transfer efficiency can be defined and the optimization design of heat exchanger can be discussed. It is found that in the irreversible processes the entransy is dissipated [117]. The more dissipation of the entransy implies the higher degree of irreversibility in heat transfer process. Thus the entransy dissipation may serve as a figure of merit for assessing the performance of heat exchanger. However, heat conduction is only one of irreversibilities occurring in heat exchange processes. For most heat exchangers, the flow friction also has a significant impact on the heat exchange performance. Fortunately, we find that the entransy is a state variable and can be employed to describe the second law of thermodynamics [4]. Therefore, the entransy can also be used for quantifying the other irreversibilities, such as the flow friction. The expressions of entransy dissipation due to heat conduction and fluid friction in heat exchanger were derived based on the second law of thermodynamics in terms of entransy dissipation [4]. When the entransy dissipation is applied to the performance evaluation and optimization design of the heat exchanger, it's necessary to be non-dimensionalised. In Ref. [8], a non-dimensionalisation method for the entransy dissipation in heat exchanger was introduced and an entransy dissipation number which can be used for evaluating the heat exchanger performance was defined.

In this subsection, the total dimensionless entransy dissipation including entransy dissipations induced by heat conduction and flow friction will been employed as the objective function to optimize the shell-and-tube heat exchanger. In addition, we shall propose the multi-objective optimization design of shell-and-tube heat exchanger which takes the entransy dissipation numbers respectively related to heat conduction and flow friction as two separate objective functions. The two optimization design approaches will be compared with each other.

7.6.2 Objective Function

As discussed in Sect. 6.3, the entransy dissipation related to heat conduction in heat exchanger can be non-dimensionalised in the following way:

$$E_{\Delta T} = \frac{\dot{G}_{diss,T}}{Q(T_{1,i} - T_{2,i})} = \frac{\dot{G}_{diss,T}}{\varepsilon(\dot{m}c)_1 (T_{1,i} - T_{2,i})^2} \tag{245}$$

which is called the entransy dissipation number related to heat conduction in heat exchanger. Similarly, the entransy dissipation related to flow friction in heat exchanger can be defined as

$$\begin{aligned} E_{\Delta P} = & \frac{-\Delta P_1}{(\rho c)_1 (T_{1,i} - T_{2,i})} \frac{1}{\ln \frac{T_{2,i}}{T_{1,i}} \left[1 - (1 - \varepsilon) \frac{T_{2,i} - T_{1,i}}{T_2} \right]} \\ & + \frac{\Delta P_2}{(\rho c)_2 (T_{1,i} - T_{2,i})} \frac{1}{\ln \frac{T_{1,i}}{T_{2,i}} \left[1 + (1 - \varepsilon) \frac{T_{2,i} - T_{1,i}}{T_{1,i}} \right]} \end{aligned} \tag{246}$$

where we assume that the working fluid in heat exchanger is incompressible. $\dot{G}^*_{diss,P}$ is called the entransy dissipation number related to flow friction in heat exchanger. The total entransy dissipation number is defined as

$$E^* = E_{\Delta T} + E_{\Delta P} \tag{247}$$

In the following, E^* will be taken as the objective function for the single-objective optimization design of shell-and-tube heat exchanger, $E_{\Delta T}$ and $E_{\Delta P}$ will be taken as two separate objective functions for the multi-objective optimization design.

7.6.3 Single-Objective Optimization

Now we take E^* as the objective function for the single-objective optimization design of shell-and-tube heat exchanger. The known data for the heat exchanger design are documented in Table 22. The working fluids on the tube and shell sides are water in our consideration. The design variables and their ranges are selected as follows:

(a) The tube outer diameter, d_o, its discrete values and the corresponding tube pitches are listed in Table 15.
(b) The whole number of heat exchange tubes, n, ranging from 50 to 550;

Table 22 The known data for heat exchanger design with the fixed heat load

	Tube side	Shell side
Inlet temperature T_i (K)	368.15	283.15
Outlet temperature T_o (K)	343.15	–
Mass flow rate $\dot{m}$ (kg/s)	50	–
Density ρ (kg/m^3)	970	991.15
Constant pressure specific heat c_p (J/kg K)	4,200	4,174
Kinematic viscosity ν (m^2/s)	3.36×10^{-7}	6.96×10^{-7}
Entrance pressure P_i (MPa)	6.5	5
Fouling resistance r (m^2 K/W)	0.000086	0.00017
Prandtl number Pr	2.015	4.5878

(c) The ratio of the baffle spacing to the shell inner diameter, *Bs*, varies between 0.2 and 1.0;
(d) The central angle of baffle cut, θ, ranging from 1.8546 to 2.9413 in radian.
(e) The outlet temperature of cold fluid, ranging from 313.15 to 343.15 K.

The constraint conditions for the heat exchanger design are:

1. Length-diameter ratio is between 6 and 10;
2. The baffle spacing is greater than 50 mm;
3. The tube side pressure drop is less than 5×10^4 Pa;
4. The shell side pressure drop is less than 5×10^4 Pa [56].

This optimization problem formulated above will be solved by the genetic algorithm. The reason for us to utilize the genetic algorithm is explained in the following. The traditional approaches to solving the optimization problems require the information of the gradients of objective functions and suffer from getting trapped at the local optimum. Thus, they can't ensure that the global optimal solution is achievable [40]. Although direct search method does not require any information about the gradient of the objective function, it depends heavily on the initial point, and frequently points to local optimum unless the objective function is unimodal [41, 122]. The genetic algorithm starts the search from a population of points; the dependence of this method on the initial point is not as strong as direct search method. Furthermore, it provides a high level of robustness by simulating nature's adaptation in the evolution process [41]. More importantly, the genetic algorithm has very strong capability to find the global optimum [42]. Therefore, the genetic algorithm [43] is employed to search the solution of the optimization problems of the heat exchanger design. The initial generation which satisfies the constraint conditions is randomly generated. The size of initial population and the maximum generation are set to 40 and 500, respectively. The variation of the best individuals' fitness value versus the number of generations is depicted in Fig. 38. It's clear that the entransy dissipation numbers due to heat conduction and flow

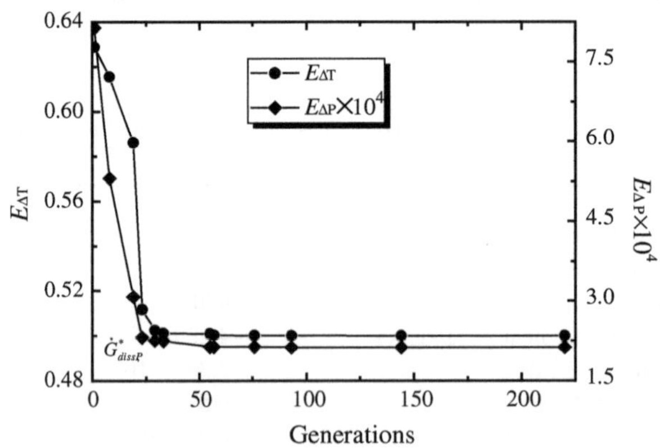

Fig. 38 Variations of $E_{\Delta T}$ and $E_{\Delta P}$ with respect to the number of generations

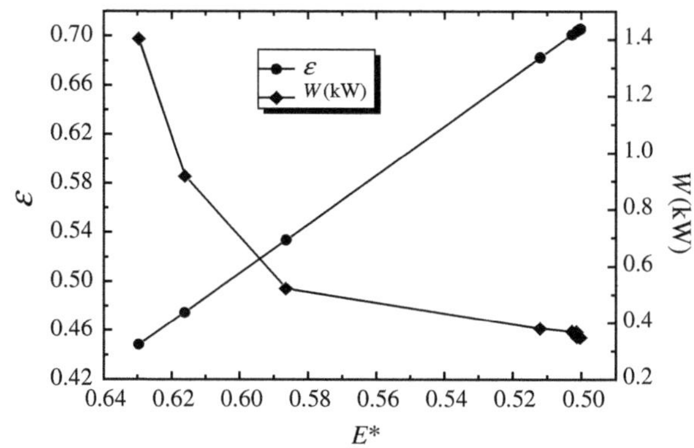

Fig. 39 Variations of the effectiveness and pumping power with total entransy dissipation number

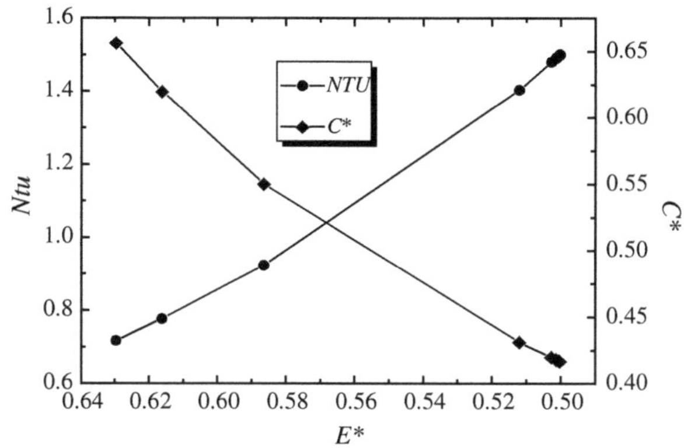

Fig. 40 Variations of *Ntu* and C^* with total entransy dissipation number

friction sharply declines firstly, and then almost keep constant beyond the 50th generation. From this figure one can see that the genetic algorithm has very high efficiency at searching the global optimal solution. Therefore, the maximum generation number which is set to 500 in the following calculation is enough to get the global optimal solution.

Figure 39 illustrates the variations of the exchanger effectiveness and pumping power with the total entransy dissipation number. Obviously, with decreasing the total entransy dissipation number, the exchanger effectiveness approximately linearly increases, while the pumping power declines sharply. Therefore, through the optimization process, the performance of heat exchanger has been improved substantially. The price paid for this improvement of heat exchanger performance is illustrated in Fig. 40. From this figure, one can see that with decreasing the total entransy dissipation number, the number of heat transfer unit (*Ntu*) increases, namely, the heat transfer area increases. Furthermore, the heat capacity flow rate ratio decreases with the decrease of the total entransy dissipation number, implying that the mass flow rate of cold fluid decreases according to the data in

Table 23 Comparison of an initial design with the optimal design

	d_o (m)	n	Bs	θ (rad)	$T_{c,o}$ (K)	NTU	C^*	W (W)	ε	$E_{\Delta T}$	$E_{\Delta P}$	E^*
Initial	0.019	243	0.977	2.038	321.26	0.717	0.656	1,403	0.448	0.63	8.14×10^{-4}	0.6296
Final	0.020	322	0.858	2.557	343.15	1.501	0.417	348	0.706	0.50	2.13×10^{-4}	0.5002

Table 24 The known data for heat exchanger design with the fixed heat transfer area

	Tube side	Shell side
Inlet temperature T_i (K)	368.15	283.15
Mass flow rate $\dot{m}$ (kg/s)	50	20
Density ρ (kg/m^3)	970	991.15
Specific heat at constant pressure c_p (J/(kg K))	4,200	4,174
Dynamic viscosity μ (kg/m s)	326×10^{-6}	690×10^{-6}
Entrance pressure P_i (MPa)	6.5	5
Fouling resistance r ((m^2 K)/W)	0.000086	0.00017
Prandtl number Pr	2.015	4.5878

Table 22 and the ranges of parameters selected in this example. In order to further demonstrate the advantages of the single-objective optimization design under the fixed heat load condition, the comparison of a randomly-generated initial design with the optimal one is shown in Table 23. From this table, it's evident that the exchanger effectiveness increases from 0.448 to 0.706, while the pumping power is reduced by 75.2 % and heat capacity rate ratio decreases from 0.656 to 0.417. Unfortunately, the number of heat transfer unit increases by about two times. So the performance of heat exchanger is improved at the expense of enlarging the heat transfer area. However, from the viewpoint of economics, it can be found that the gross profit is far more than the increase of the investment cost, the detailed analysis is presented in Ref. [58].

When more attention is paid to Table 23, it can be found that the entransy dissipation number due to flow friction is around three orders of magnitude less than that caused by heat conduction. In fact, the irreversibility due to fluid friction is far less than the irreversibility associated with heat conduction for liquid–liquid heat exchanger [34]. Hence, the single-objective optimization of heat exchanger design which takes the total entransy dissipation number as the objective function may ignore the influence of flow friction. This can be demonstrated by the heat exchanger design with the fixed heat transfer area.

The known data for the heat exchanger design is presented in Table 24, the total entransy dissipation number is taken as the objective function, the design parameters and their ranges are the same as that presented in the last example, except the outlet temperature of the cold fluid. The heat transfer area is fixed at 60 m^2, the sizes of initial population and the maximum number of generations are set to 40 and 500, respectively. The genetic algorithm is employed to solve this optimization problem. The variations of $E^*_{\Delta T}$ and $E^*_{\Delta P}$ with respect to the number of generations are shown in Fig. 41. From this figure, it's evident that with increasing the number of generations, the entransy dissipation number due to heat conduction

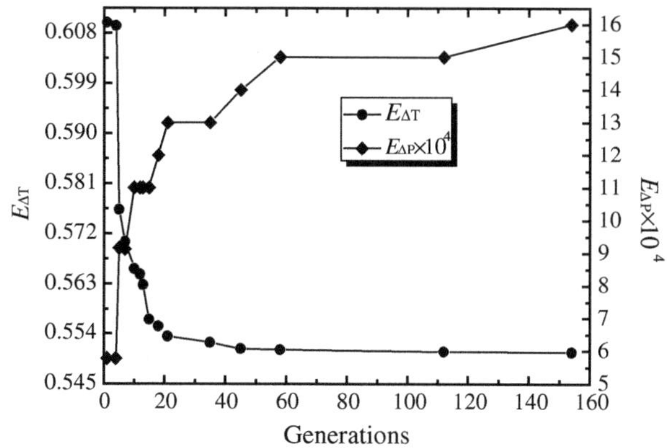

Fig. 41 Variations of $E_{\Delta T}$ and $E_{\Delta P}$ with respect to the number of generations

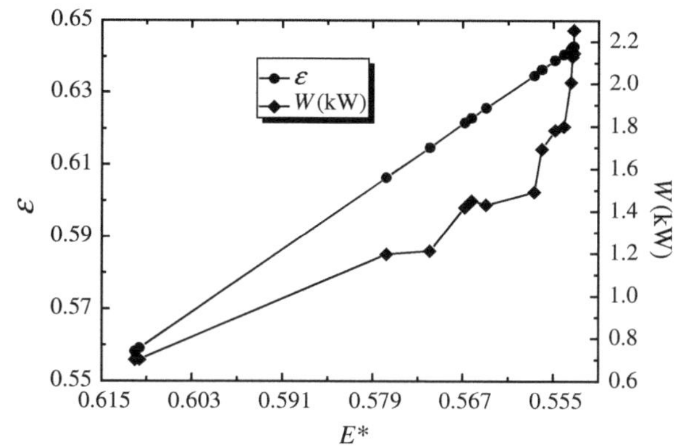

Fig. 42 Variations of the exchanger effectiveness and pumping power with the total entransy dissipation number

decreases remarkably, while the entransy dissipation number caused by flow friction rises significantly, which is undesirable. The variations of the exchanger effectiveness and pumping power with respect to the total entransy dissipation number are illustrated in Fig. 42. Clearly, with decreasing the total entransy dissipation number, the exchanger effectiveness is improved significantly, while the pumping power increases drastically. Recall that the heat transfer area is fixed in this example, thus the improvement of the exchanger effectiveness is at the expense of the larger pumping power consumption. From Figs. 41 and 42, one can see that taking the total entransy dissipation number as the objective function is almost equivalent to minimizing the entransy dissipation number only due to heat conduction, and the entransy dissipation caused by flow friction is almost neglected since it's far smaller than that caused by heat conduction. In an attempt to solve this problem, the multi-objective optimization design of heat exchanger is established in the following subsection.

7.6.4 Multi-objective Optimization

Mathematically, the multi-objective optimization minimizes several objectives simultaneously, with a number of inequality or equality constraints. For details about the multi-objective optimization, please refer to Sect. 3.2.2. In the following the entransy dissipation numbers respectively caused by heat conduction and flow friction are taken as two separate objective functions. The design parameters, their bounds and the constraints remain the same as that specified in the single-objective optimization design case under the given heat load condition. The known data of the heat exchanger is shown in Table 22. In addition, the population type is *double vector*, and the size of initial population is set as 40. *Feasible population operator* is selected to create a well-dispersed initial population that satisfies all bounds and constraints of design parameters. The algorithm for choosing parents from individuals is *tournament*. Set *crossover fraction* to 0.8 so that 80 % of the next generation is produced by *crossover operator*. *Scattered crossover* function is used to combine two individuals, or parents, to form a child for the next generation. *Gaussian Mutation* function is used, the values of *scale* and *shrink* parameters are fixed as one, which means that the standard deviation shrinks linearly from 1 to 0 as the last generation is reached. The direction of *migration* is forward, i.e., the nth subpopulation migrates into the $(n + 1)$th subpopulation, and *migration* takes place every 20 generations. A controlled elitist genetic algorithm (a variant of NSGA-II [31]) is adopted for searching the optimal solutions, which can help increase the diversity of the population even if they have lower fitness values. The diversity of population is controlled by the elite members of the population in the process; the *distancecrowding* function helps to maintain diversity by favoring individuals that are relatively far away on the front. The *Pareto fraction* is set to 0.35 so as to limit the number of individuals in the current population that are on the *Pareto front* to 35 % of the population size. The total number of generations is set to 500, which serves as the stopping criteria to terminate the iterative process.

Some representative optimal solutions obtained by the multi-objective optimization are listed in Table 25. From this table, one can see that the larger effectiveness corresponds to the smaller pumping power. The comparison between Tables 23 and 25 shows that the optimal solution in Table 25 which has the same exchanger effectiveness as that in Table 23 requires less pumping power consumption. The required pumping power is reduced by 22.4 % through the multi-objective optimization process. Therefore, the multi-objective optimization demonstrates obvious advantages over the single-objective optimization.

For the second example of heat exchanger optimization design with the given heat transfer area, the known data for heat exchanger design is documented in Table 24, the design variables and their ranges remain the same as specified in the single-objective optimization design of heat exchanger with the fixed the heat transfer area. But the entransy dissipation numbers caused by heat conduction and flow friction are taken as two separate objectives in the optimization design. The Pareto front obtained by the multi-objective optimization is shown in Fig. 43. Figure 43a illustrates the variations of the heat conduction and flow friction

Table 25 The Pareto front obtained by multi-objective optimization design with the fixed heat load

	d_o (m)	n	Bs	θ(rad)	$T_{c,o}$ (K)	K_o (W/m^2 K)	C^*	W (W)	ε	$E_{\Delta T}$	$E_{\Delta P}$	E^*
Multi-objective	0.022	283	1.00	2.362	69.14	1,260.5	0.423	260.2	0.696	0.505	1.58×10^{-4}	0.5052
	0.022	283	1.00	2.349	69.26	1,263.1	0.422	260.2	0.697	0.504	1.58×10^{-4}	0.5045
	0.022	283	1.00	2.320	69.53	1,268.6	0.420	260.2	0.700	0.503	1.58×10^{-4}	0.5029
	0.022	283	1.00	2.299	69.71	1,272.5	0.419	260.3	0.703	0.502	1.58×10^{-4}	0.5019
	0.022	283	0.95	2.295	69.86	1,276.6	0.418	265.0	0.704	0.501	1.61×10^{-4}	0.5010
	0.022	283	0.91	2.294	70.00	1,280.2	0.417	270.2	0.706	0.500	1.64×10^{-4}	0.5002

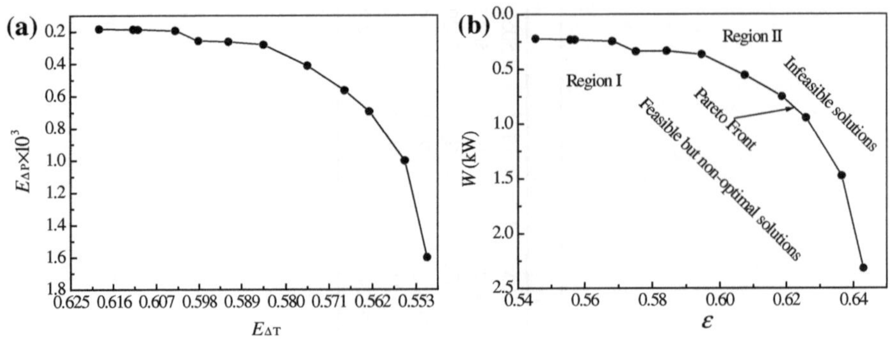

Fig. 43 The Pareto front for heat exchanger with the fixed heat transfer area: **a** $E_{\Delta T}$ and $E_{\Delta P}$; **b** the pumping power versus the exchanger effectiveness

entransy dissipation numbers for different optimal solutions in the Pareto optimal set. The pumping power and the exchanger effectiveness corresponding to the optimal solutions are depicted in Fig. 43b. In this figure, there are two regions which are formed by the Pareto front. The solutions in region I are feasible but not the optimal solutions, while ones in region II represent the infeasible solutions. Note that a set of optimal solutions are available for the multi-objective optimization approach, therefore, it provides more alternatives for heat exchanger design than the single-objective optimization approach.

According to the entransy dissipation theory, the entransy dissipation can be used for describing the irreversibilities induced by heat conduction and flow friction. Such irreversibilities are the main factors to deteriorate the performance of heat exchanger. Therefore, in this subsection, based on the entransy dissipation theory and genetic algorithm, two optimization approaches for heat exchanger design are proposed. Firstly, a single-objective optimization approach is formulated, where the total entransy dissipation number is taken as the objective function. When the heat load is fixed, the single-objective optimization design can significantly improve the performance of heat exchanger. However, for the case that the heat transfer area is fixed, the improvement of exchanger effectiveness through the optimization process is at the expense of the increase of the pumping power. In order to address this problem, the multi-objective optimization design of heat exchanger is established, where the entransy dissipation numbers respectively related to heat conduction and flow friction are taken as two separate objectives. In comparison with the single-objective optimization approach, the multi-objective optimization design of heat exchanger can achieve the same exchanger effectiveness with less consumption of pumping power. Furthermore, the multi-objective optimization leads to the non-unique optimal solutions which provide more flexibility for the heat exchanger design.

References

1. Guo, J.F., Xu, M.T., Cheng, L.: Multi-objective optimization of heat exchanger design by entropy generation minimization. ASME J. Heat Transf. **132**, 081801 (2010)
2. Guo, J.F., Cheng, L., Xu, M.T.: Optimization design of shell-and-tube heat exchanger by entropy generation minimization and genetic algorithm. Appl. Therm. Eng. **29**, 2954–2960 (2009)
3. Guo, J.F., Cheng, L., Xu, M.T.: The entropy generation minimization based on the revised entropy generation number. Int. J. Exergy **7**(5), 607–626 (2010)
4. Xu, M.T.: The thermodynamic basis of entransy and entransy dissipation. Energy **36**, 4272–4277 (2011)
5. Xu, M.T., Guo, J.F., Cheng, L.: Application of entransy dissipation theory in heat convection. Front. Energy Power Eng. Chin. **3**(4), 402–405 (2009)
6. Xu, M.T.: Variational principles in terms of entransy for heat transfer. Submitted to Energy (2011)
7. Guo, J.F., Xu, M.T., Cheng, L.: Principle of equipartition of entransy dissipation for heat exchanger design. Sci. China Technol. Sci. **53**(5), 1309–1314 (2010)
8. Guo, J.F., Cheng, L., Xu, M.T.: Entransy dissipation number and its application in heat exchanger performance evaluation. Chin. Sci. Bull. **54**(15), 2708–2713 (2009)
9. Guo, J.F., Xu, M.T., Cheng, L.: The application of entransy dissipation theory in optimization design of heat exchanger. In: Proceedings of the Fourteenth International Heat Transfer Conference, Washington, DC, USA, 8–13 Aug 2020
10. Guo, J.F., Xu, M.T., Cheng, L.: The entransy dissipation minimization principle under given heat duty and heat transfer area conditions. Chin. Sci. Bull. **56**(19), 2071–2076 (2011)
11. Li, X.F., Guo, J.F., Xu, M.T., Cheng, L.: Entransy dissipation minimization for optimization of heat exchanger design. Chin. Sci. Bull. **56**(20), 2174–2178 (2011)
12. Bejan, A.: Entropy generation through heat and fluid flow. Wiley, New York (1982)
13. Bejan, A.: Entropy generation minimization. CRC Press, New York (1995)
14. Bejan, A.: Entropy generation minimization: the new thermodynamics of finite-size devices and finite-time processes. J. Appl. Phys. **79**(3), 1191–1218 (1996)
15. Bejan, A.: The thermodynamic design of heat and mass transfer processes and devices. Int. J. Heat Fluid Flow **8**(4), 258–276 (1987)
16. Bejan, A.: The concept of irreversibility in heat exchanger design: counter-flow heat exchangers for gas-to-gas applications. ASME J. Heat Transf. **99**, 374–380 (1977)
17. Yilmaz, M., Sara, O.N., Karsli, S.: Performance evaluation criteria for heat exchangers based on second law analysis. Exergy Int. J. **1**(4), 278–294 (2001)
18. Bejan, A.: A study of entropy generation in fundamental convective heat transfer. ASME J. Heat Transf. **101**(4), 718–725 (1979)
19. Sekulic, D.P.: The second law quality of energy transformation in a heat exchanger. ASME J. Heat Transf. **112**(2), 295–300 (1990)
20. Witte, L.C., Shamsundar, N.: A thermodynamic efficiency concept for heat exchange devices. J. Eng. Power-Trans. ASME **105**, 199–203 (1983)
21. Sciubba, E.: A minimum entropy generation procedure for the discrete pseudo-optimization of finned-tube heat exchangers. Rev. Gen. Therm. **35**(416), 517–525 (1996)
22. Bejan, A.: General criteria for rating heat exchanger performance. Int. J. Mass Heat Transf. **21**(5), 655–658 (1978)
23. Grazzini, G., Gori, F.: Entropy parameters for heat exchanger design. Int. J. Heat Mass Transf. **31**(12), 2547–2554 (1988)
24. Zhang, L.W., Balachandar, S., Tafti, D.K., Najjar, F.M.: Heat transfer enhancement mechanisms in in-line and staggered parallel-plate fin heat exchangers. Int. J. Heat Mass Transf. **40**(10), 2307–2325 (1997)
25. Ordonez, J.C., Bejan, A.: Entropy generation minimization in parallel-plates counterflow heat exchangers. Int. J. Energy Res. **24**(10), 843–864 (2000)

26. Bejan, A.: Thermodynamic optimization of geometry in engineering flow systems. Exergy, Int. J. **1**(4), 269–277 (2001)
27. Bejan, A.: Fundamentals of exergy analysis, entropy generation minimization, and the generation of flow architecture. Int. J. Energy Res. **26**(7), 545–565 (2002)
28. Poulikakos, D., Bejan, A.: Fin geometry for minimum entropy generation in forced convection. ASME J. Heat Transf. **104**(4), 616–623 (1982)
29. Vargas, J.V.C., Bejan, A., Siems, D.L.: Integrative thermodynamic optimization of the crossflow heat exchanger for an aircraft environmental control system. ASME J. Heat Transf. **123**(4), 760–769 (2001)
30. Oğulatu, R., Doba, F., Yilmaz, T.: Irreversibility analysis of cross flow heat exchangers. Energy Convers. Manage. **41**(15), 1585–1599 (2000)
31. Reddy, B.V., Ramkiran, G., Kumar, K.A., Nag, P.K.: Second law analysis of a waste heat recovery steam generator. Int. J. Heat Mass Transf. **45**(9), 1807–1814 (2002)
32. Sahiti, N., Krasniqi, F., Fejzullahu, X.H., Bunjaku, J., Muriqi, A.: Entropy generation minimization of a double-pipe pin fin heat exchanger. Appl. Therm. Eng. **28**(17–18), 2337–2344 (2008)
33. Mishra, M., Das, P.K., Sarangi, S.: Second law based optimisation of crossflow plate-fin heat exchanger design using genetic algorithm. Appl. Therm. Eng. **29**, 2983–2989 (2009)
34. Hesselgreaves, J.E.: Rationalisation of second law analysis of heat exchanger. Int. J. Heat Mass Transf. **43**(22), 4189–4204 (2000)
35. London, A.L., Shah, R.K.: Costs of irreversibilities in heat exchanger design. Heat Transf. Eng. **4**, 59–73 (1983)
36. Kuppan, T.: Heat Exchanger Design Handbook. Marcel Dekker Inc, New York (2000)
37. Shah, R.K., Sekulic, D.P.: Fundamentals of Heat Exchanger Design. John Willey, Hoboken (2003)
38. Kays, W.M., London, A.L.: Compact Heat Exchanger, 3rd edn. McGraw-Hill, New York (1984)
39. Qian, S.W.: Heat Exchanger Design Handbook. Chemical industrial Press, Beijing (2002). (in Chinese)
40. Babu, B.V., Munawar, S.A.: Differential evolution strategies for optimal design of shell-and-tube heat exchangers. Chem. Eng. Sci. **62**, 3720–3739 (2007)
41. Oh, Y.H., Kim, T., Jung, H.K.: Optimal design of electric machine using genetic algorithm coupled with direct method. IEEE Trans. Mag. **35**(3), 1742–1744 (1999)
42. Fanni, A., Marchesi, M., Serri, A., Usai, M.: A greedy genetic algorithm for continuous variables electromagnetic optimization problems. IEEE Trans. Mag. **33**(2), 1900–1903 (1997)
43. Houck, C.R., Joines, J., Key, M.: A genetic algorithm for function optimization: a Matlab implementation. ACM Trans. Math. Softw. (1996)
44. Goldberg, D.E.: Genetic Algorithms I Search, Optimization and Machine Learning. Addison-Wesley, Reading (1989)
45. Holland, J.H.: Adaptation in Nature and Artificial System. The University of Michigan Press (1975)
46. Caputo, A.C., Pelagagge, P.M., Salini, P.: Heat exchanger design based on economic optimization. Appl. Therm. Eng. **28**(10), 1151–1159 (2008)
47. Yun, J.Y., Lee, K.S.: Influence of design parameters on the heat transfer and flow friction characteristics of the heat exchanger with slit fins. Int. J. Heat Mass Transf. **43**(14), 2529–2539 (2000)
48. Xu, Z.M., Yang, S.R., Chen, Z.Q.: A modified entropy generation number for heat exchagners. J. Therm. Sci. **5**(4), 257–263 (1996)
49. Xie, G.N., Sunden, B., Wang, Q.W.: Optimization of compact heat exchangers by a genetic algorithm. Appl. Therm. Eng. **28**, 895–906 (2008)
50. Selbas, R., Kızılkan, Ö., Reppich, M.: A new design approach for shell-and-tube heat exchangers using genetic algorithms from economic point of view. Chem. Eng. Process. **45**, 268–275 (2006)

51. Mishra, M., Das, P.K., Saranqi, S.: Optimum design of crossflow plate-fin heat exchangers through genetic algorithm. Int. J. Heat Exch. **5**, 379–401 (2004)
52. Zhou, M., Sun, S.: Genetic Algorithms: Theory and Application. National Defence Industry Press, Beijing (1999). (In Chinese)
53. Shi, M.Z., Wang, Z.Z.: Principles and Design of Heat Exchanger Device. Southeast university press, Nanjing (1996). (In Chinese)
54. Palen, J.W.: Heat Exchanger Sourcebook. Hemisphere Publishing Cor, Washington (1986)
55. Bell, K.J., Mueller, A.C.: Wolverine Engineering Data Book II. Wolverine Tube Inc. (2004)
56. State bureau of quality and technical supervision: Tubular heat exchangers, GB151-1999. Standards Press of China, Beijing (1999) (in Chinese)
57. Bejan, A.: Advanced Engineering Thermodynamics. Wiley, New York (1988)
58. Guo, J., Xu, M., Cheng, L.: The application of field synergy number in shell-and-tube heat exchanger. Appl. Energy **86**, 2079–2087 (2009)
59. Prigogine, I.: Introduction to Thermodynamics of Irreversible Processes, 3rd edn. Wiley, New York (1967)
60. Vargas, J.V.C., Bejan, A.: Thermodynamic optimization of finned crossflow heat exchanger for an aircraft environmental control system. Int. J. Heat Fluid Flow **22**, 657–665 (2001)
61. Bejan, A.: Second law analysis in heat transfer. Energy **5**, 721–762 (1980)
62. Raznjevic, K.: Handbook of Thermodynamic Tables, 2nd edn. Begell House Inc., New York (1995)
63. Hilbert, R., Janiga, G., Baron, R., Thévenin, D.: Multi-objective shape optimization of a heat exchanger using parallel genetic algorithms. Int. J. Heat Mass Transf. **49**, 2567–2577 (2006)
64. Barakat, T.M.M., Fraga, E.S., Sørensen, E.: Multi-objective optimization of batch separation process. Chem. Eng. Process. **47**(12), 2303–2314 (2008)
65. Deb, K.: Multi-Objective Optimization using Evolutionary Algorithms. Wiley, Chichester, England (2001)
66. Copiello, D., Fabbri, G.: Multi-objective genetic optimization of the heat transfer from longitudinal wavy fins. Int. J. Heat Mass Transf. **52**, 1167–1176 (2009)
67. Kondepudi, D., Prigogine, I.: Modern Thermodynamics-From Heat Engines to Dissipative Structures. Wiley, Chichester (1998)
68. de Nevers, N., Seader, J.D.: Lost work: a measure of thermodynamic efficiency. Energy **5**, 757–769 (1980)
69. Szargut, J.: International progress in second law analysis. Energy **5**, 709–718 (1980)
70. Anern, J.E.: Applications of the second law of thermodynamics to cryogenics-A review. Energy **5**, 891–897 (1980)
71. Berg, C.A.: Process integration and the second law of thermodynamics: future possibilities. Energy **5**, 733–743 (1980)
72. Kotas, T.J.: The exergy method of thermal plant analysis. Butterworths, London (1985)
73. Bejan, A., Tsatsaronis, G., Moran, M.: Thermal design and optimization. Wiley, New York (1996)
74. Bizarro, J.P.S.: Entropy production in irreversible processes with friction. Phys. Rev. E **78**, 021137 (2008)
75. Ben-Amotz, D., Honig, J.M.: Average entropy dissipation in irreversible mesoscopic processes. Phys. Rev. Lett. **96**, 020602 (2006)
76. Ben-Amotz, D., Honig, J.M.: Rectification of thermodynamic inequalities. J. Chem. Phys. **118**, 5932 (2003)
77. Curzon, F., Ahlborn, B.: Efficiency of a Carnot engine at maximum power output. Am. J. Phys. **43**, 22 (1975)
78. Sohrab, S.H.: A scale-invariant model of statistical mechanics and modified forms of the first and the second laws of thermodynamics. Int. J. Therm. Sci. **38**, 845–853 (1999)
79. den Broek, C.V.: Thermodynamic efficiency at maximum power. Phys. Rev. Lett. **95**, 190602 (2005)

80. Esposito, M., Lindenberg, K.: Universality of efficiency at maximum power. Phys. Rev. Lett. **102**, 130602 (2009)
81. Esposito, M., Kawai, R., Lindenberg, K., der Broeck, C.V.: Efficiency at maximum power of low-dissipation Carnot engines. Phys. Rev. Lett. **105**, 150603 (2010)
82. Clausius, R.: The mechanical theory of heat-with its applications to the steam engine and to physical properties of bodies. John van Voorst, 1 Paternoster Row, MDCCCLXVII, London (1865)
83. Herwig, H.: The role of entropy generation in momentum and heat transfer. In: Proceedings of the 14th International Heat Transfer Conference, Paper No. IHTC14-23348, Washington DC, USA, 8–13 Aug 2010
84. Bertola, V., Cafaro, E.: A critical analysis of the minimum entropy production theorem and its application to heat and fluid flow. Int. J. Heat Mass Transf. **51**, 1907–1912 (2008)
85. Shah, R.K., Skiepko, T.: Entropy generation extrema and their relationship with heat exchanger effectiveness-number of transfer unit behavior for complex flow arrangement. ASME J. Heat Transf. **126**, 994 (2004)
86. Guo, Z.Y., Zhu, H.Y., Liang, X.G.: Entransy—A physical quantity describing heat transfer ability. Int. J. Heat Mass Transf. **50**, 2545 (2007)
87. Guo, Z.Y., Liu, X.B., Tao, W.Q., Shah, R.K.: Effectiveness-thermal resistance method for heat exchanger design and analysis. Int. J. Heat Mass Transf. **53**, 2877 (2010)
88. Liu, X., Wang, M., Meng, J., Ben-Naim, E., Guo, Z.Y.: Minimum entransy dissipation principle for the optimization of transport networks. Int. J. Non-Linear Sci. Numer. Simul. **11**(2), 113–120 (2010)
89. Chen, Q., Wang, M., Pan, N., Guo, Z.Y.: Optimization principles for convective heat transfer. Energy **34**(9), 1199–1206 (2009)
90. Chen, Q., Kang, Y., Wang, M., Pan, N., Guo, Z.Y.: A new approach to analysis and optimization of evaporative cooling system I: theory. Energy **35**(6), 2448–2454 (2010)
91. Cheng, X., Liang, X.: Entransy flux of thermal radiation and its application to enclosures with opaque surfaces. Int. J. Heat Mass Transf. **54**, 269 (2011)
92. Chen, L., Wei, S., Sun, F.: Constructal entransy dissipation rate minimization of a disc. Int. J. Heat Mass Transf. **54**, 210 (2011)
93. Xu, M.T.: Variational principles in terms of entransy for heat transfer. Energy **33**(1), 973 (2011)
94. Onsager, L.: Reciprocal relations in irreversible process. I. Phys. Rev. **37**, 405–426 (1931)
95. Onsager, L., Machlup, S.: Fluctuations and irreversible processes. Phys. Rev. **91**, 1505–1512 (1953)
96. Attard, P.: Statistical mechanical theory for steady state systems. II. Reciprocal relations and the second entropy. J. Chem. Phys. **122**, 154101 (2005)
97. Attard, P.: Statistical mechanical theory for steady state systems. VI. Variational principles. J. Chem. Phys. **125**, 214502 (2006)
98. Bejan, A.: Shape and Structure: From Engineering to Nature. Cambridge University Press, Cambridge (2000)
99. Hunt, K.L.C., Hunt, P.M.: Deviations from minimum entropy production at steady states of reacting chemical systems arbitrarily close to equilibrium. Physica A **154**, 207–211 (1988)
100. Tykodi, R.J.: Thermodynamics of steady states: a weak entropy-production principle. Physica **72**, 341–354 (1974)
101. Hillert, M., Ågren, J.: Extremum principles for irreversible processes. Acta Mater. **54**, 2063–2066 (2006)
102. Barragan, D.: Entropy production and Newton's cooling law. Rev. Ing. E Invest. **29**, 88–93 (2009)
103. Attard, P.: Statistical mechanical theory for steady state systems. VI. Variational principles. J. Chem. Phys. **125**, 214502 (2006)
104. Evans, D.J., Cohen, E.G.D., Morriss, G.P.: Probability of second law violations in shearing steady states. Phys. Rev. Lett. **71**, 2401–2404 (1993)

105. Yilmaz, M., Sara, O.N., Karsli, S.: Performance evaluation criteria for heat exchangers based on second law analysis. Exergy, Int. J. **1**, 278–294 (2001)
106. Wang, S.P., Chen, Q.L., Zhang, B.J.: An equation of entransy transfer and its application. Chin. Sci. Bull. **54**, 3572–3578 (2009)
107. Chen, Q., Ren, J.X.: Generalized thermal resistance for convective heat transfer and its relation to entransy dissipation. Chin. Sci. Bull. **53**, 3753–3761 (2008)
108. Xia, S.J., Chen, L.G., Sun, F.R.: Optimization for entransy dissipation minimizatino in heat exchanger. Chin. Sci. Bull. **54**, 3587–3595 (2009)
109. Liu, X.B., Meng, J.A., Guo, Z.Y.: Entropy generation extremum and entransy dissipation extremum for heat exchanger optimization. Chin. Sci. Bull. **54**, 943–947 (2009)
110. Wu, J., Liang, X.G.: Application of entransy dissipation extremum principle in radiative heat transfer optimization. Sci. China Ser. E: Technol. Sci. **51**, 1306–1314 (2008)
111. Hausen, H.: Heat Transfer in Counter Flow, Parallel Flow and Cross Flow. McGraw-Hill Inc., New York (1983)
112. Tondeur, D., Kvaalen, E.: Equipartition of entropy production: an optimality criterion for transfer and separation process. Ind. Eng. Chem. Res. **26**, 50–56 (1987)
113. Sauar, E., Ratkje, S.K., Lien, K.M.: Equipartition of forces: a new principle for process design and optimization. Ind. Eng. Chem. Res. **35**, 4147–41533 (1996)
114. Balkan, F.: Comparison of entropy minimization principles in heat exchange and a short-cut principle: EoTD. Int. J. Energy Res. **27**, 1003–1014 (2003)
115. Guo, Z.Y., Li, Z.X., Zhou, S.Q., Daxi, X.: Principle of uniformity of temperature difference field in heat exchanger. Sci. China Tech. Sci. **39**(1), 68–75 (1996)
116. Guo, Z.Y., Cheng, X.G., Xia, Z.Z.: Least dissipation principle of heat transport potential and its application in heat conduction optimization. Chin. Sci. Bull. **48**(4), 406–410 (2003)
117. Han, G.Z., Guo, Z.Y.: Physical mechanism of heat conduction ability dissipation and its analytical expression. Proc. CSEE **27**, 98–102 (2007)
118. Wei, S.H., Chen, L.G., Sun, F.R.: "Volume–Point" heat conduction constructal optimization with entransy dissipation minimization objective based on rectangular element. Sci. China Tech. Sci. **51**(8), 1283–1295 (2008)
119. Wei, S.H., Chen, L.G., Sun, F.R.: Constructal multidisciplinary optimization of electromagnet based on entransy dissipation minimization. Sci. China Tech. Sci. **52**(10), 2981–2989 (2009)
120. Johannessen, E., Nummedal, L., Kjelstrup, S.: Minimizing the entropy production in heat exchanger. Int. J. Heat Mass Transf. **45**, 2649–2654 (2002)
121. Bedeaux, D., Standaert, F., Hemmes, K., Kjelstrup, S.: Optimization of process by equipartition. J. Non Equilibr. Thermodyn. **24**, 242–259 (1999)
122. Mahammed, O.A., Uler, G.F.: A hybrid technique for the optimal design of electromagnetic devices using direct search and genetic algorithms. IEEE Trans. Mag. **33**, 1931–1934 (1997)

Author Index

L. Wang (ed.), *Advances in Transport Phenomena 2011*,
Advances in Transport Phenomena 3, DOI: 10.1007/978-3-319-01793-8,

MIX
Papier aus verantwortungsvollen Quellen
Paper from responsible sources
FSC® C105338

If you have any concerns about our products,
you can contact us on
ProductSafety@springernature.com

In case Publisher is established outside the EU,
the EU authorized representative is:
Springer Nature Customer Service Center GmbH
Europaplatz 3, 69115 Heidelberg, Germany

Printed by Libri Plureos GmbH
in Hamburg, Germany